How to Fossilize Your Hamster

NewScientist

How to Fossilize Your Hamster

and other amazing experiments for the armchair scientist

Mick O'Hare

PENGUIN CANADA

Published by the Penguin Group

Penguin Group (Canada), 90 Eglinton Avenue East, Suite 700, Toronto, Ontario, Canada M4P 2Y3 (a division of Pearson Canada Inc.)

Penguin Group (USA) Inc., 375 Hudson Street, New York, New York 10014, U.S.A.
Penguin Books Ltd, 80 Strand, London WC2R 0RL, England
Penguin Ireland, 25 St Stephen's Green, Dublin 2, Ireland (a division of Penguin Books Ltd)
Penguin Group (Australia), 250 Camberwell Road, Camberwell, Victoria 3124, Australia (a division of Pearson Australia Group Pty Ltd)
Penguin Books India Pvt Ltd, 11 Community Centre, Panchsheel Park, New Delhi – 110 017, India
Penguin Group (NZ), 67 Apollo Drive, Rosedale, North Shore 0632, New Zealand (a division of Pearson New Zealand Ltd)
Penguin Books (South Africa) (Pty) Ltd, 24 Sturdee Avenue, Rosebank, Johannesburg 2196, South Africa

Penguin Books Ltd, Registered Offices: 80 Strand, London WC2R 0RL, England

Published in Canada by Penguin Group (Canada), a division of Pearson Canada Inc., 2008
Published in the U.K. by Profile Books Ltd, 2007

1 2 3 4 5 6 7 8 9 10 (WEB)

Text illustrations by Frazer Hudson
Margin illustrations by Brett Ryder

Manufactured in Canada.

Library and Archives Canada Cataloguing in Publication data available upon request

British Library Cataloguing in Publication data available

Visit the Penguin Group (Canada) website at **www.penguin.ca**

Special and corporate bulk purchase rates available; please see **www.penguin.ca/corporatesales** or call 1-800-810-3104, ext. 477 or 474

Contents

Introduction

Experiments are what make science tick. Observing, recording and observing again have taught us everything we know about our universe and the world around us. Without experimental evidence, science is reduced to little more than a set of theories. Real scientists – from Newton and the apple to Pavlov and his dogs – get out there and observe or, in the case of readers of *New Scientist* magazine, they go into the kitchen or garden and prove just why and how things happen. And when scientists have observed and recorded what they have seen, they repeat the experiment to ensure its results are verifiable.

Which is what this book is all about – experimenting and seeing for yourself. It's where the real pleasure in science lies. Science isn't boffins in labs, it's people experimenting wherever they are and using whatever is to hand. By reading this book you'll understand how great science has been achieved through experimentation.

In true *New Scientist* fashion, we focus on the trivial. You can find out why shaken and stirred martinis taste different, but not why the universe is expanding. You'll learn how to extract iron from your breakfast cereal, but you won't discover what's inside a black hole. The big stuff can come later, after you've been inspired by the experiments here. Best of all, you'll be able to ensure that your dear departed hamster is preserved for eternity (though you won't be able to discover just how long eternity is).

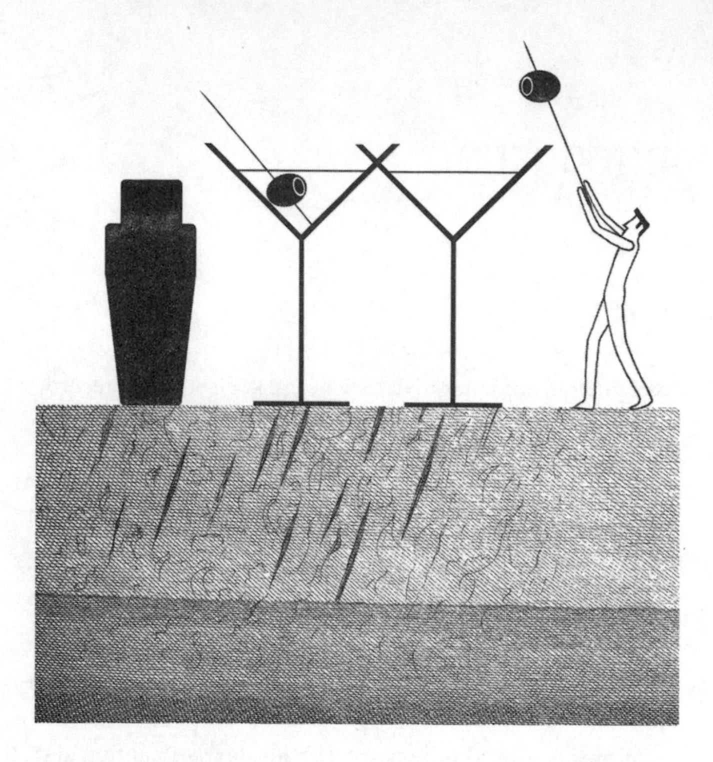

While not every entry can be termed an experiment, they all allow you to try out simple things at home that demonstrate science in action. These range from the chemical (why do cola and Mentos make such an explosive mix?) to the biological (why do some things pass straight through the digestive system apparently untouched?). And we've unearthed a preponderance of experiments featuring alcoholic drinks – *New Scientist* is staffed by journalists after all. While these ones are for adults only, the rest are aimed at the entire family. Children should be supervised when carrying out any of the experiments we describe. For although they have been chosen with safety in mind, a few involve potentially hazardous materials (hot water, matches, knives…), so care needs to be taken.

Some of the experiments will almost certainly need a large outdoor space to try out, but you should be able to attempt most of them as soon as you get this book home or shortly after a trip to the local shops. Hardly any require specialist equipment. If you need further instruction on how to make them work, you'll find some of them on our website – www.newscientist.com/hamster.

The best science is always collaborative, so I am grateful to all the readers of *New Scientist*, colleagues, scientists and smart people everywhere who have helped make this book possible – a roll call appears on pages 197–200.

Finally, remember: scientists' theories are there to be challenged. If you carry out the experiments in this book and draw different conclusions, let us know – you will find our contact details in the Acknowledgements. Science is constantly evolving in response to new evidence; that is what makes it so wonderful and enlightening. There's nothing we'd enjoy more than being proved wrong by someone who has carried out an experiment from this book and arrived at a different conclusion. After all, real scientists always experiment, not once, not twice, but again and again …

Mick O'Hare

1 In the Living Room

Cream on

Why does a 2 mm layer of cream on Tia Maria create rapidly circulating and swirling patterns?

This is a perfect experiment for the after-dinner tabletop laboratory, and as it's one that *New Scientist* brought to the attention of the world, we feel particularly smug. Use a good, single pouring cream – milk is too runny, double cream too thick. You can even drink the end product.

What do I need?

- a wide-bottomed glass
- single cream
- a bottle of Tia Maria
- a spoon

What do I do? Pour 1 cm of Tia Maria into the glass and add a very thin layer of single cream, poured gently over the back of the spoon so that it spreads over the surface of the Tia Maria.

What will I see? The surface of the cream will break up into a pattern of cells as it reacts with the Tia Maria.

What's going on? The swirling patterns are caused by convection – the bulk movement of fluids usually caused by temperature differences. However, in this case the convection is driven by a difference in concentration between the two liquids. This is called solutal convection.

The alcohol in the Tia Maria is the driving force. After the cream is added, the alcohol begins to diffuse through it. When it reaches the surface it alters the surface tension, reducing it. Areas unaffected maintain their higher surface tension and so pull the low-surface tension liquid areas towards them. As the surface liquid is pulled away, Tia Maria moves up to fill its place. This has even less surface tension than the liquid pulled away and the process becomes self-sustaining as a convection cycle is created. This will last as long as a difference in concentration between the cream and the Tia Maria remains.

Surface tension convection is known as Marangoni convection. It plays a part in drying different paints and causes patterns such as those seen in wine 'legs' running down the inside of a drinking glass (see 'Drinker's legs' experiment on page 13). Similar effects are seen in rolling clouds or oil frying in a pan. Tia Maria is something of an oddity though, because in other systems the convection cells are circular or hexagonal.

PS: We are especially pleased to include this experiment because when the question was first asked in *New Scientist* nobody knew the answer, which led to a group of researchers studying the phenomenon and writing the definitive paper on the subject.

Want to read more? The paper can be read in *Physica A*, vol. 314, p. 291.

Over the top

Sparkling wine or beer poured into a dry glass froths up spectacularly, but if the inside of the glass is wet, it doesn't. Why?

Quite frankly, we are amazed at the number of home experiments you can do that involve alcoholic drinks. If you are only interested in experimenting, you don't have to drink the wine or beer, but ...

What do I need?

- freshly opened sparkling wine and/or beer
- champagne flutes and/or beer glasses
- olive oil
- a tablespoon of sugar
- dust (look behind the wine rack)

What do I do? Pour some of the wine or beer rapidly into a vertical glass so that it froths up to the rim, let the bubbles subside and then add more.

What will I see? The second batch of liquid will not have the same frothing effect as the first.

What's going on? Beer, sparkling wine and other fizzy drinks are supersaturated with gas. Although the laws of thermodynamics favour the gas bubbling out of the dissolved state, bubble formation is unlikely since bubbles must start small. Because the pressure of these tiny bubbles can reach about 30 atmospheres in a bubble only 0.1 micrometres in diameter and the solubility of gases increases with increasing pressure (Henry's law), the gas is forced back into solution as quickly as it comes out.

However, bubbles can form around dust particles, surface irregularities and scratches. These areas, know as nucleation sites, are hydrophobic (they repel water) and allow gas pockets to form without first forming the tiny bubbles. Once the gas pocket reaches a critical size, it bulges out and rounds up into a properly convex bubble, the radius of curvature of which is sufficiently large to prevent the self-collapse described earlier.

Then there is a cascade effect. If the bubbles reach a certain critical number per unit volume, this in itself constitutes a physical disturbance and results in the release of yet more bubbles.

Nucleation may be precipitated by a variety of imperfections. Minute salt crystals, such as calcium sulphate, may be present if the glass has been left to dry by evaporation after being washed in hard water. Or there may be tiny cotton fibres if the glass has been dried with a tea cloth. Dust particles may have settled on the glass if it has been left standing upright on a shelf for any length of time. And tiny scratches will be present on the inside surface of all but brand-new glasses.

Once the inside of the glass is wet, any salt crystals will have dissolved and any cotton fibres will no longer function as centres of nucleation. Most of the dust particles and all of the scratches will, of course, still be there. However, these will have been coated with liquid and the fresh carbonated liquid will reach them very slowly, by diffusion. Bubbles will still be produced, but at a rate that is too slow for the cascade effect to come into play. As a result, the drink will not froth over.

To demonstrate this, take a glass and thoroughly coat the inside with olive oil, which is a more efficient surface-covering agent than water. Then add the wine or beer (if you are feeling frugal, use a carbonated drink such as lemonade). The effervescence will be nil or minimal. Throw in the dust you scraped from behind the wine rack and see how it can affect the behaviour of the liquid. Finally, add a few million centres of

nucleation from a large spoonful of granulated sugar and prove to yourself how volcanic the effervescence can be.

PS: Thanks to modern production techniques, today's glasses are of such good quality that some manufacturers build in deliberate imperfections, especially in beer glasses, in order to generate enough bubbles to maintain the head on the top of your tipple.

There are almost certainly similarities between this experiment and the Mentos and cola effect from the 'Overreaction' experiment on page 194, although the jury is still out on why Mentos and cola produce quite such a spectacular effect. If you are going to do either of these indoors, go for the wine or beer version ...

White water drinking

Why do anisette-based drinks, such as Pernod, ouzo or sambuca, turn white when water is added?

This is an interesting effect because two completely colourless liquids combine to produce an opaque white one, immediately transforming something that was in solution and invisible into something insoluble and visible.

What do I need?

- drinks glasses
- a bottle of pastis, ouzo or sambuca
- a jug of water

What do I do? Pour one of the three anisette-based drinks or, if you wish to be a true scientist running controls alongside your main experiment (see the 'Fizz fallacy' experiment on page 28), a glass of all three. Add water.

What will I see? The drinks will turn from clear to milky white or, in the case of pastis, milky yellow.

What's going on? Anisette-based drinks rely on aromatic compounds called terpenes for their flavour. These are soluble in alcohol, but not in water. The 40 per cent or so alcohol by volume present in the drinks is enough to keep the terpenes dissolved, but when water is added they are forced out of solution as the alcohol content of the drink falls, to produce a milky-looking suspension.

Absinthe, a similar drink based on wormwood, gives a more impressive green suspension. Terpenes are responsible for a lot of the harsher plant scents and flavours found in the drink, including those of lemon grass and thyme.

PS: Absinthe, always legal in the UK, but rarely drunk for many decades due to a paucity of producers, is now undergoing a revival. However, dealers need a special licence to sell it and it remains an almost mythical spirit reputed – but not proven – to have psychotropic effects. However, its toxicity means that it is still illegal in other parts of the world, including the USA, where the Food and Drug Administration states that the presence of the chemical thujone in absinthe has harmful neurological effects, although absinthe producers claim thujone levels are insufficient to induce any response in humans. Interestingly, pastis was first produced by the distillers Pernod and Ricard after absinthe was banned in France in 1915. They removed the wormwood and added more star anise to produce one of the first versions of the drink you've been experimenting on.

You only drink twice

Is there any difference between a vodka martini that is shaken, not stirred?

James Bond is famously fastidious in demanding that his martinis are only ever served shaken. Apparently, there *is* a difference…

What do I need?

- a decent recipe for a vodka martini. The following one, in which olives and citrus fruit are added prior to shaking or stirring, comes from www.drinkoftheweek.com
- 140 ml of vodka
- eight drops of vermouth
- two olives and twists of lime or lemon
- cocktail shakers
- martini glasses

What do I do? Mix the ingredients in a jug (gently, to avoid any undue shaking or stirring). Carefully divide the mixture into two cocktail shakers. Shake one, stir the other, then pour the martini into two separate glasses.

What will I see? Not a lot, except two exquisite cocktails, but you may – just may – be able to taste the difference. Ideally you'll need to recruit a volunteer for a blind tasting – don't tell them which is which.

What's going on? Supposedly, when a martini is shaken not stirred it 'bruises' the vodka spirit in the drink. To seasoned martini drinkers this alters the taste.

However, there is great debate as to what is really happening. It is impossible to bruise alcoholic spirit in the way that you can bruise fruit (or indeed your arm) because it is a

liquid and has no vascular system, although the olive and citrus fruit (if you add them to the cocktail mix before shaking or stirring as our recipe does) can be bruised by the shaking action, thus releasing flavoursome oils and juice. Other recipes, however, eschew adding olives and fruit beforehand so that only the vodka and vermouth are shaken or stirred, thus ruling out botanical intervention in the flavour. We suggest you repeat this experiment adding the olives and fruit after shaking or stirring and see if you can tell the difference.

Nonetheless, martini drinkers can still detect a difference when the vodka and vermouth are shaken or stirred without the olives and fruit. This is because a martini is usually drunk within seconds of preparation, rather than minutes. The tiny bubbles caused by the shaking mean that a well-shaken martini is cloudy. This has an effect on the texture of the drink – it is less oily than the stirred version – and hence the taste is ever so slightly altered.

The bubbles from shaking can also partially oxidise the aldehydes in the vermouth in the way that the flavour of red wine alters when we oxidise it – commonly known as letting the wine 'breathe'. This can alter the flavour of the martini – again, ever so slightly.

So now that you know, be honest – can you tell the difference between shaken and stirred?

PS: Agent 007 has provided *New Scientist* with plenty of material over the years. One of the most surprising discoveries we made was that the kind of silencer that Bond uses when ruthlessly dispatching the bad guys is more myth than fact. It seems Hollywood has taken great liberties with silencers and most real ones are much larger than the cigar-tube sized versions shown in the Bond films, and are very difficult to fit and remove. And if James Bond did manage to fit a silencer quickly enough, his gun would not produce the distinctive 'phut' we hear in the films – more of a firm crack, similar to the sound of a car door being slammed. For, despite what you see in the movies, it is almost impossible to silence a gun because the gap between the cylinder and the barrel allows gas to escape. Sadly, because this is a book of experiments that you can try at home, we cannot recommend a trial of this particular 007 phenomenon. We include it merely on a whim to add to the ambience of your vodka martini experiment.

Drinker's legs

What causes the thin, separated films that run down the inside of a glass containing whisky or other spirits?

These are known as 'legs' and are often considered the measure of a good spirit. You can be the judge of this if you try a variety of spirits and check out whether the good ones show the phenomenon best.

What do I need?

- a glass tumbler (or several)
- whisky (and other spirits if you wish to compare and contrast)
- water

What do I do? Pour a decent measure of whisky and swill it round the glass. Do the same with a glass of water of the same volume. Take sips of both and watch the liquids drain back into the bottom of the glasses down the sides. You can try other spirits and see if they show a difference.

What will I see? The thin film of whisky draining back into the glass after you have swirled it and sipped it will separate into rivulets, which will be roughly equally spaced. These are the 'legs' mentioned above. Water, however, drains back more rapidly, 'legs' do not develop and the most you will see on the inside of the glass is a few beads of liquid.

What's going on? This phenomenon was first correctly interpreted by James Thomson, the elder brother of Lord Kelvin (best known for his scale of absolute temperature measurement and his work on thermodynamics). In an 1855 paper entitled 'On certain curious Motions observable at the Surfaces of Wine and other Alcoholic Liquors', published in *Philosophical Magazine* (vol. 10, p. 330), Thomson explained that the effect is caused by capillary motion, or surface tension, explained in both 'Flower power' and the 'Floaters' experiments on pages 21 and 58.

Alcoholic drinks are mixtures of ethanol and water, from which the ethanol – which has a lower boiling point than water – evaporates faster than the water when the drink has been poured. The surface tension of ethanol is lower than that of water and the different rates of evaporation induce conflicting forces between the more watery regions of the drink and the more alcoholic regions. In effect, the watery and alcoholic regions are pulled apart, leaving legs of increasingly watery liquid as the ethanol evaporates and the watery remains start to trickle back down the glass under their own weight.

Thomson's paper won him little fame. This kind of surface tension convection is now known as Marangoni convection after Carlo Marangoni, who worked on similar studies 20 years later. (See the 'Cream on' experiment on page 5.)

Marangoni convection can cause other remarkable effects. For example, if the whisky is placed in a very flat dish and exposed to the air, the whole body of liquid becomes divided into hexagonal columns that circulate as the alcohol evaporates. Unfortunately, this is difficult to see in the home lab.

Normal convection in liquids, caused by density differences, usually swamps the effect of Marangoni convection and it cannot get to work except in special circumstances like the side of your whisky glass. However, in the weightless conditions of space, Marangoni convection rules supreme and is therefore a fertile topic for research by space scientists.

PS: 'Legs' in liquors and wines have been noted for millennia. The first account may even have appeared in the Bible: 'Look not thou upon the wine when it is red, when it giveth his colour in the cup, when it moveth itself aright' (Proverbs 23:31). It seems that even then the effect was considered an indicator of good-quality alcohol.

Yo ho ho

Why does dark rum form a frothy scum on the top of the drink when mixed with cola, but white rum does not?

As you have now discovered, adults should never turn down the chance to experiment with alcohol. Here's a chance to try one of the world's most popular drinks, the Cuba libre, while engaging in some serious scientific endeavour.

What do I need?

- a bottle of dark rum
- a bottle of white rum
- cola
- lots of tumblers (all good experiments are repeated to ensure consistency of results)
- a good detergent

What do I do? Pour a decent measure of white rum into one tumbler and a similar amount of dark rum into another. Top up the glasses with cola (the less you use, the more dense the resulting scum will be).

What will I see? On the surface of the drink in the glass with dark rum and cola you'll see a lacy network of darkish scum. Consume some of the drink and you'll see that the scum lines the glass in the way that a stale head from beer leaves a residue on the inside of your glass.

What's going on? Dark rum is essentially white rum plus colourings and flavourings. One of the traditional flavourings used in dark rum is molasses – the residue left after the juice of sugar cane has been boiled, concentrated and had as much sugar as possible extracted from it. Molasses is a good source of calcium and magnesium salts; it also contains tarry caramelised compounds plus degraded organic substances such as waxes, fatty acids and polysaccharides.

Cola (flavoured solutions of phosphoric acid), when combined with the rum's solutions of alkaline earth ions such as calcium and magnesium, forms insoluble phosphates. In water you would probably see these as clouds, but in low concentrations in a mix such as rum, tiny crystals and bubbles are bound into a scum by the fatty acids and waxes. This readily congeals into the material you see on the surface of the drink.

PS: Remember to wash your Cuba libre glasses straight after the experiment. The scum is notoriously difficult to remove if you leave it until the next day and your mood will not be enhanced if you consumed too many during your research.

Beer orders

If you are making a shandy and the beer is poured into the glass first, followed by the lemonade, the contents will fizz up and possibly overflow. But add the beer to the lemonade and this doesn't happen? Why?

This is a good ice-breaker at parties or in the pub. Of course, as all true aficionados know, the best shandy is made with ginger beer not lemonade. Kids can't drink the shandy, of course, but they can have fun pouring and then consuming the dregs of the soft drinks afterwards. They can also check out the claim that lemonade makes you burp more than ginger beer does.

What do I need?

- tall beer glasses
- bottles of beer
- lemonade and ginger beer
- paper towels for mopping up

What do I do? Take two of the glasses and half-fill them with beer. Add lemonade to one and ginger beer to the other. Repeat the experiment with two more glasses but put the soft drinks in first and then top them both up with beer.

What will I see? The glasses half-filled with beer first will froth up as the lemonade or ginger beer is added, but the effect will be much less marked if the glasses are half-filled with the soft drinks first and then topped up with beer.

What's going on? If you take another three clean glasses and add only beer, lemonade or ginger beer to each, you'll notice that beer is the only drink to form a head. This is because it contains substances known as surfactants as well as proteins

and other long-chain molecules that help it to form and stabilise bubbles. Lemonade, on the other hand, forms bubbles that fizz and burst very quickly. Ginger beer lies somewhere between the two.

When you pour lemonade into beer it rapidly sinks, hits the bottom, bounces back and starts to bubble. Because these bubbles pass up through the beer above they pick up the surfactants in the beer and form a froth on the surface. Beer poured into lemonade sinks quickly in the same way, but the bubbles that push up from the bottom in this case pass through the lemonade, which contains no surfactants, and pop quickly when they reach the surface. By the time enough beer has reached the top, the drink has settled down and little fizz is left.

PS: Does lemonade make you burp more than ginger beer? Well, lemonade has more carbon dioxide added to it than ginger beer, but ginger beer gives you that fizzy feeling in your nose if you drink it too quickly. Which you find more distasteful is probably a matter of personal preference. The consensus is that beer probably tastes better than both.

Ouch!

What causes the pain induced by a piece of aluminium foil on a tooth filling?

If you have looked after your teeth well, you can smugly sit this one out. Those with fillings, however, will quickly become enlightened ...

What do I need?

- aluminium foil
- saliva
- a tooth with an amalgam filling (still in situ)

What do I do? Generate a decent amount of saliva in your mouth, pop the aluminium foil in and position it over a filling in your teeth, preferably a molar so you can bite on the foil.

What will I feel? You may jump! You'll feel anything from a slight tingling in the filled tooth to actual pain as the foil touches it.

What's going on? When two dissimilar metals are separated by a conducting liquid, a current will flow between them, and this can stimulate nerves. In this case the two dissimilar metals are the amalgam in your tooth and the aluminium foil. A thin film of saliva separates the foil from the filling and, because it is a reasonable electrolyte containing various salts, it acts as the conducting liquid allowing a current to flow between tooth and filling. Because the filling is close to the dental nerve, the current will stimulate it, causing pain.

PS: Luigi Galvani first discovered the dissimilar metal/electrolyte effect in 1762, when he carried out experiments using the nerves in frogs' legs. When probes of differing metallic composition were applied to the frogs, their legs twitched. Using tooth amalgam and spit is a more humane way of carrying out this experiment, although the owner of the filling might not agree.

Shocking TV

Why, when you switch off your television in a darkened room and touch the screen, is there a crackle of sound and a fluorescent glow at the point you touch the screen?

This is more fun than yet another episode of reality TV and hints at what's really inside your TV – other than Elvis, of

course. You'll also feel the hairs on your arm stand up (again, nothing to do with Elvis).

What do I need?

- a non-flat-screen, old-fashioned TV (hopefully you kept yours in the loft or have given it to the kids)
- a darkened room
- a finger

What do I do? Switch off the lights, turn off the TV and touch the screen – don't wait too long or you won't see the effect.

What will I see? You'll see a fluorescent glow at the point where you touch the otherwise blank screen and you'll hear a crackle of static electricity discharge. Move your finger or hand across the screen and hear more static as the hairs on your arm simultaneously become charged in the electrical field of the TV.

What's going on? An old-fashioned TV has a cathode-ray tube which creates a picture by passing electricity from the cathode (the negative electrode by which electrons enter an electrical device) at the rear of the TV's vacuum tube to the anode (the positive electrode) at the front. (Modern, liquid crystal display screens create the TV picture using a different process.) The anode on an old-style cathode-ray tube is created by a thin layer of aluminium coating the back of the TV screen, and pictures are created by fluorescence acting on the phosphor, the synthetic fluorescent that also coats this screen. When the TV is switched on this anode induces a charge on the outside of the screen. When the TV is switched off, most of this charge leaks away, but some lingers on the phosphor. This charge flows when it is touched by your finger and the phosphor glows in the way it would if the screen was still switched on, while the crackle is the sound of the electric discharge.

Flower power

How do flowers 'drink' water?

If you forget to tend to flowers in a vase, you'll discover the astonishing rate at which they take up water. A full vase can become empty in no time at all, leaving you with drooping stems and fallen petals. So where does all that water go?

What do I need?

- white flowers, such as lilies or carnations
- food colouring (red or blue give the best effect) or some ink
- water
- scissors or a knife
- a clear glass vase
- a sunny window ledge

You can also use celery if you want to study this effect further.

What do I do? Trim each flower stem with the scissors or knife and place them in the vase containing the water plus the food colouring or ink. Make sure the colour in the vase is distinct – the darker the better – but also make sure the water does not become thick if you are using ink; it should still be 'watery'. Place the vase in a sunny spot. Check the flowers every hour.

What will I see? Over the course of a few hours the petals will change colour. The patterns that form can be quite splendid, with vein-like structures across the petal surfaces and petal edges that are deeply coloured.

What's going on? Plants need water in order to live and grow, and they do this by drawing up water from the soil

Slice through a celery stick to give a newly cut surface and place the stick in the vase containing the coloured water.

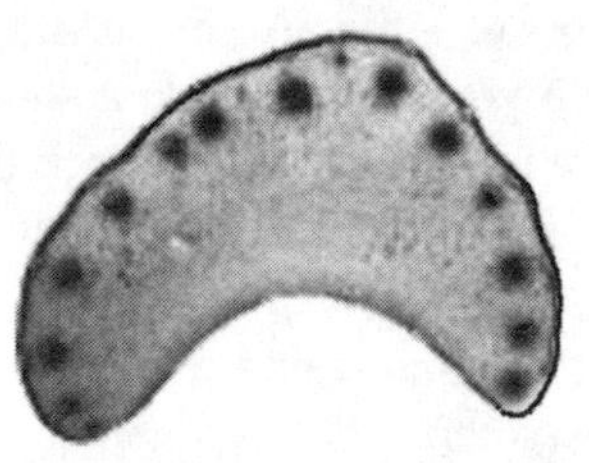

After a few hours, remove it and slice through the stick again to reveal the xylem structure

through their roots (or, in this case, from the water in the vase). The process is driven by transpiration, which is the evaporation of water from petal and leaf surfaces (the French word for 'to sweat' is *transpirer*). When the leaf surfaces become deficient in water more is drawn from the vase as water molecules are pulled upwards to replace those which have been transpired. Obviously, a sunny, warm spot speeds up this process. Normally all this is invisible, but the ink or food colouring demonstrates exactly what is going on.

The coloured water is drawn through vascular structures (the xylem) in the plant. At its simplest level xylem acts like a straw. As water is transpired – or 'sucked' – away from the petal or leaf surface more water is pulled into the straw to replace it. Xylem conducts water and dissolved nutrients upwards from the root. At their most dense, the hollow channels of xylem form the internal scaffolding of woody stems as they branch out through leaves and petals. You can see the structure of xylem if you remove one of the flowers from the vase and cut through the stem. Vein-like tubes will be clearly visible in the cross-section, filled with coloured liquid.

Over the course of about 10 hours you'll see the spread of the coloured water throughout your flowers, until they are entirely dyed. Plants, especially in hot conditions, can take up vast amounts of water, which is why a vase that was full one morning can be almost empty the next. On a hot day, a full-grown tree, such as a birch or sycamore, can draw upwards of 500 litres from the ground, in some cases aided by root pressure, forcing water upwards into the xylem; all of which is later transpired through evaporation.

PS: Celery sticks show up xylem structure better than the thin stems of flowers. Remove a celery stick from a new bunch, slice off the bottom end to give a newly cut surface and place the stick in the vase containing the coloured water. Leave it for a few hours before returning and removing the stick. Wipe it clean and cut through the stick – you'll see distinct dots on the cut surface indicating where xylem has taken up the coloured water.

Shedding light

Why does the flame on a candle that is standing on a revolving turntable point inwards and not outwards?

When we first heard of this phenomenon we were amazed. But when we attempted to replicate the experiment in the *New Scientist* office we couldn't get it to work. When the candle was lit and the turntable switched on, the flame trailed behind the candle as it orbited the centre of the turntable in the way it does when you walk along holding a candle. Then somebody suggested a jam jar…

What do I need?

- a candle
- a turntable (a record player turntable set to 45 rpm – or 78 rpm if you have a very old gramophone – will do, although a potter's wheel works just as well, as does a rotating cheeseboard)
- a large jar or similar

If you want to take the experiment further you'll need:

- a car
- a helium balloon
- a spirit level

What do I do? Attach the candle firmly to the turntable. A heavy base of plasticine helps, but a wide-based, heavy candle holder is even better. Light the candle. Place the upturned jar over the candle and switch on the turntable. Make sure that air can get under the base of the jar or the candle will go out when it becomes starved of oxygen. You might combine this experiment with the 'Air space' experiment on page 99 because much of the equipment is the same.

What will I see? The candle flame will not be thrown outwards by the motion of the revolving turntable, nor will it trail behind the candle, thanks to the jar protecting it from any breeze. It will actually point inwards towards the centre of the turntable.

What's going on? The air in the rotating jar is, in effect, being spun in a centrifuge. This means the candle flame bends towards the inside of the turntable for the same reason that it normally points straight upwards: the heated gas that comprises the flame is less dense than the cooler surrounding air. So it is this denser, cooler air that moves outwards when the turntable is rotating and the heated, less dense gas of the flame that moves inwards, as centripetal force (the force that acts on a body moving in a circular path and is directed toward the centre around which the body is moving) acts on the rotating system.

We tend to think that the flame is composed of 'something' because we can see it, and that the surrounding air space is composed of 'nothing' because we can't see it. Therefore the brain intuitively suspects that the flame will move outwards as the turntable rotates, in the way that a piece of cotton or tissue paper would. But cotton and tissue paper are denser than the surrounding air, whereas a flame is less dense. Therefore the visible flame moves inwards while the invisible air moves outwards, apparently in contradiction of our brain's interpretation of visible objects.

Another, perhaps simpler, way of understanding why the candle flame points inwards is to consider a similar problem. If you are driving a car which contains a helium balloon held by a string and you brake hard, you will slam forward against the seat belt but the balloon will move to the back of the car.

This is because the air in the car has inertia and continues forward just as you do, while the balloon reacts by floating towards the area of lowest pressure – the lowest-density portion of the air mass, which is found at the back of the car. Similarly, the tethered balloon will lean forwards under acceleration and towards the inside of bends as the car turns, in the way the candle leans inwards as it rotates.

The candle flame, like the balloon, is buoyant, its shape resulting from a complex interaction between the burning wax at the wick and the heating of the surrounding air. So, like the balloon, the flame also floats in the direction of lowest pressure – towards the axis of rotation at the centre of the turntable. To complete the comparison, the candle, like the car, is accelerated with respect to the air surrounding the flame, so the air is moving radially outwards relative to the candle. The flame reacts by floating inwards.

If you do attempt this further extension of the experiment it should go without saying that the driver should keep his or her eyes on the road, not attempt heavy braking if the conditions are slippery, and the experiment should be carried out by a passenger who will not distract the driver. It should only be tried on private land where there are no other vehicles or pedestrians, and all occupants of the vehicle must wear correctly fitted seat belts.

As a final visual test of this phenomenon, place a spirit level on your turntable, pointing away from the centre like a bicycle wheel spoke and switch the turntable on. The bubble moves inwards not outwards, because the spirit inside the instrument is denser than the air bubble and consequently the bubble is pushed inwards.

PS: Sue Ann Bowling of the University of Alaska tells us that if she was being really picky she would argue that the less dense candle flame is accelerated more by the centripetal force acting on the air in the jar. Newton's law says that for the same force, the product of mass and acceleration is the same. So if the mass is smaller, the acceleration must be more. Her explanation is, of course, technically correct, but for us lay scientists it is simpler to comprehend that the centripetal force has more effect on the denser air than it does on the less dense flame.

2 In the Kitchen

Fizz fallacy

I've been told that champagne will keep its fizz overnight if a teaspoon is suspended in the neck of the bottle. How does this work?

We have included this not because it is spectacular or even because it actually works. It is here to emphasise the importance of having controlled conditions when attempting to prove or disprove a theory.

What do I need?

- a few bottles of champagne
- a fridge
- a teaspoon
- some champagne flutes

What do I do? Open two bottles of champagne. Drink some from each, then place a teaspoon into the neck of one of the bottles, with the handle dangling downwards and with no part of the teaspoon touching the liquid. Drink a little more if it is. The other bottle should be left open. To maintain a true control, try to keep the amount of champagne in each bottle equal. Now place both in the fridge and leave them overnight. Test them at regular intervals, noting how fizzy each bottle is and whether there is any significant difference between the two. We suggest you test them the following morning, lunchtime and evening, and again on subsequent days until the champagne in both bottles has lost all its fizz.

What will I see? You'll need to be objective in your assessment of the fizz quotient of each bottle, but you will find, especially if you repeat the experiment as any decent scientist would, that both bottles are equally fizzy at each stage of testing. As an objective measure you could see how much champagne needs to be added to the glass for the bubbles to reach the rim, but for this to work you'll have to make sure you pour all glasses at the same rate.

What's going on? This question and its subsequent elevation to the status of urban myth is a classic example of being misled by uncontrolled experiments. People think this experiment works because in the morning their half-bottle of champagne with its teaspoon in the neck is still fizzy. The truth of the matter is that champagne – surprisingly – keeps its fizz for days with or without the teaspoon, as your tasting experiments will show.

The opened bottle without the teaspoon acts as a control against which you can gauge the fizziness of the bottle with the teaspoon in its neck. Both bottles decrease in fizziness at exactly the same rate.

Because people rarely have two bottles open at the same time and have stored an unfinished bottle with a suspended spoon in its neck they have attributed the following day's unexpected longevity of fizz to the spoon. However, you will have found by now that champagne stays fizzy for three days or more.

It's not uncommon to attach significance to apparently linked events when there is no control data to compare them with. Often you'll hear people saying things like, 'How incredible, I was just thinking of you when the phone rang and it was you…' Telepathy is not involved here. We merely

ignore the huge number of times we think of people and the phone did not ring.

PS: If your budget doesn't stretch to champagne, the experiment will work just as well with Cava, Asti Spumante or a domestic sparkling wine.

Pasta puzzle

If a strand of dried spaghetti is held at both ends and bent, why does it always break into three or more pieces?

This is, indeed, a strange phenomenon. Surely holding a strand of dried spaghetti at both ends and bending it until it breaks should produce just two pieces, but it hardly ever does – usually three or even more pieces are the result. This conundrum first appeared in *New Scientist* in 1995 and was repeated in 1998. Even so, we didn't get to the bottom of it until 2006. It's a problem that has taxed greater minds than ours, including that of the Nobel prize-winning physicist Richard Feynman.

What do I need?

- strands of dried spaghetti
- something to catch them in

What do I do? Hold a strand of spaghetti at both ends and bend it until it breaks. Repeat with the other strands.

What will I see? In nearly all cases the spaghetti will break into three or more pieces. Even on the rare occasions when it seems to break into only two you'll often find a stray shard or splinter has flown off into the nether regions of your kitchen.

What's going on? Back in 1998 a *New Scientist* reader tried to get to grips with the issues involved and came very close to solving the problem.

First, when you bend a piece of spaghetti, it does not usually break at the apex of the bend where the stresses are highest, because failure in the spaghetti is controlled by defects in the pasta. The first break occurs at a point near the apex where the combination of stress level and defect size reaches a critical value. This breaks the original piece into a long and a short piece. After the break, as the longer piece snaps back, the whipping action sends the tip beyond the neutral point (the original straight state of the piece of spaghetti) and activates the next defect on what was the long side. This defect has already been opened up on the outside of the curved spaghetti by the first bending, so it doesn't take much to finish off the crack by bending it in the other direction.

Second, the sequence of events can be determined by looking at the broken ends of the spaghetti pieces. When a break occurs, the fracture starts cleanly on the stretched, convex side and ends slightly raggedly on the compressed, concave side where a small splinter – or spicule – is usually torn away from one side of the break. Additionally, careful inspection of the ejected middle piece of spaghetti will reveal evidence of spicule formation at both ends and that these are on opposite sides. This shows that the two breaks which generate the middle piece occur while the spaghetti is bending in opposite directions, which is consistent with the dynamics of linear spaghetti structures.

While being very much on the right track, the reader's observations only partly answered the conundrum. It took Basile Audoly and Sebastien Neukirch to verify what was going on in their paper 'Fragmentation of rods by cascading cracks: Why spaghetti does not break in half', published in *Physical Review Letters* (vol. 95, p. 95505), which won them the 2006 Ig Nobel Prize for Physics (see below).

①
spaghetti

②
Snap
both strands subject to flexural waves after snapping

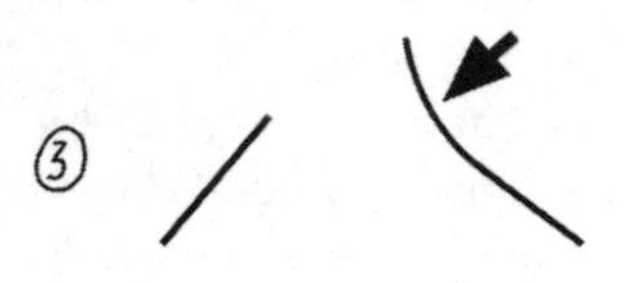
③
flexural wave causes spaghetti to spring back after breaking, already weakened here by previous bending.

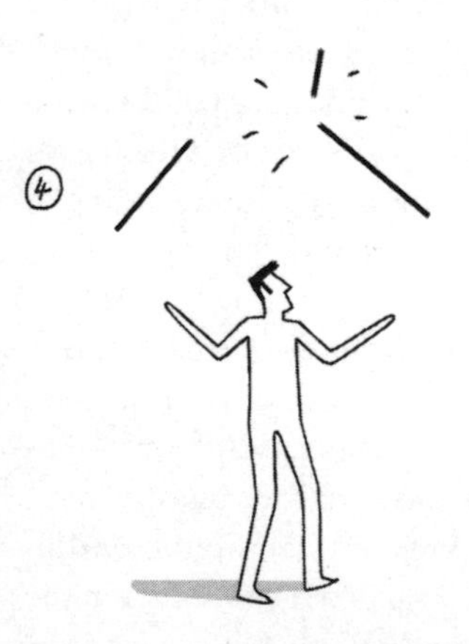
④
third piece snaps off after springing back.

Audoly and Neukirch broke strands of spaghetti of varying thicknesses and lengths by clamping one end and bending them from the other. They found that the unexpected three-part breakage occurs because of what are known as flexural waves. When the curvature of the spaghetti reaches a critical point, the first break appears. The shock of this causes a flexural wave to ripple down each of the two resulting lengths of pasta at high speed and amplitude.

The two halves formed by the initial break do not have time to relax and straighten before being hit by the flexural wave, which causes them to curve even further and suffer more breaks, leading to a cascade of cracks in the pasta. Often more than three pieces are created when this happens.

While spaghetti snapping is in itself a rather humdrum if fun pastime, Audoly and Neukirch's work also provides important information about failures in other elongated, brittle structures, including human bones and bridge spans.

PS: In *No Ordinary Genius,* the illustrated biography of Richard Feynman published in 1994, Danny Hills describes his and Feynman's experiments with spaghetti: 'If you get a spaghetti stick and break it, it turns out that instead of breaking in half, it will almost always break into three pieces. Why is this true – why does it break into three pieces? ... Well we ended up at the end of a couple of hours with broken spaghetti all over the kitchen and no real good theory about why spaghetti breaks into three.' This seems to have been a common occurrence – apparently visitors to Feynman's home were often presented with sticks of spaghetti and asked to help solve the problem.

There is, therefore, a delicious irony in the fact that while this puzzle drove Feynman, a physics Nobel prize winner, to distraction, those who discovered the reason why it happened were awarded the antithesis of Nobel fame, the Ig Nobel Prize for Physics in 2006, 41 years after Feynman won his Nobel prize. Nobel prizes are awarded for supreme achievements in scientists' chosen fields while Ig Nobels, from the opposite end of the research spectrum, are awarded for success in the areas of improbable research, humour and, quite often, silliness.

Want to read more? A video of breaking spaghetti can be seen at www.lmm.jussieu.fr/spaghetti/index.html, where you can also find out more about Audoly and Neukirch's research.

No Ordinary Genius is edited by Christopher Sykes and published by W. W. Norton and Company.

Banana armour

Does banana skin turn brown faster in a fridge than out of it?

This is counterintuitive to those of us brought up to believe that chilling foodstuffs slows decay, but a simple experiment will show us if it's true or not.

What do I need?

- two or more bananas (and possibly some fresh banana skins)
- a fridge (and possibly a domestic freezer)
- lemon juice

What do I do? Place one banana in the fridge and leave the other at room temperature (approx. 20 °C). Observe each banana three or four times a day and note the relative discoloration of the skins. As a side-experiment, rub a third banana with lemon juice before subjecting it to the fridge conditions.

What will I see? The banana in the fridge will brown or blacken faster than the one at room temperature. However, a banana rubbed with lemon juice and placed in the fridge will not decay at the same rate as the untreated one.

What's going on? While many fruits are stabilised by refrigeration, most tropical and subtropical fruits, and bananas in particular, exhibit chill injury. Tests show that the ideal temperature for bananas is 13.3 °C. Below 10 °C spoilage is accelerated because their cells' internal membranes are damaged, releasing enzymes and other substances. Banana skin can blacken overnight as it softens and breaks down.

The membranes that keep separate the contents of the various compartments inside a cell are essentially two layers of slippery fat molecules or lipids. Chill these membranes and the molecules get more sticky, making the membranes less flexible. Bananas adjust the composition of their membranes to give the right degree of membrane fluidity for the temperature at which they normally grow. They do this by varying the amount of unsaturated fatty acids in the membrane lipids: the greater the level of unsaturated fatty acid, the more fluid the membrane at a given temperature. If you chill the fruit too much, areas of the membrane become too viscous and it loses its ability to keep the different cellular compartments separate. Enzymes and substrates that are normally kept apart therefore mix as the membranes collapse and hasten the softening of the fruit flesh.

Skin blackening involves the action of a different enzyme from those involved in flesh softening. In the skin, polyphenol oxidase breaks down naturally occurring phenols in the banana skin into substances similar in structure to the melanin found in suntanned human skin. So the browning starts sooner in refrigerated bananas because of chill-induced membrane damage that allows the normal process of decay – which would have occurred at room temperature – to begin earlier. The cold itself does not speed up the browning part of the reaction. Indeed, if chill damage in a fridge is induced first, removing the banana from the fridge then hastens the process

as the reaction that causes the browning, once it is under way, is speeded up by heat.

This can be demonstrated by putting a banana skin in a freezer for a few hours. The inner surface will stay creamy white because, although the membranes are destroyed by the freezing process, the oxidases cannot work at such low temperatures. Then let it thaw overnight at room temperature. In the morning it will be pitch-black due to the damage the membranes suffered in the freezer. Yet had the cold itself caused the blackening, it would have turned dark while it was being frozen.

Decay can be slowed by acids, which prevent the release of the polyphenol oxidase enzyme. This is why adding lemon juice – which is rich in citric acid – to skins can slow the browning process. A similar slowing of the breakdown process can be seen if bananas are coated in wax; this stops oxygen reaching the skin and speeding up decay.

PS: In commercial storage of tropical fruit, chilling injury is a big problem. Unlike temperate fruits such as apples and pears, which can be stored at temperatures close to freezing, tropical fruits break down faster, which is contrary to our usual experience of refrigeration. Because tomatoes, now commonplace in northern Europe and North America, are actually a semi-tropical fruit, evidence suggests that they will last longer out of the fridge than in it. We look forward to hearing about readers' home experiments on this and any number of other fruits whose growing environments straddle temperate and tropical climes.

Iron man

Breakfast cereals often claim to be fortified with iron. Well, are they?

They are and, more amazingly, if you have a magnet you can extract it too! So ponder the ingredients list on your packet of cornflakes while you are munching breakfast, and then set about removing one of them…

What do I need?

- breakfast cereal fortified with iron (cornflakes work, but check on the side of the packet to see what the iron content is – the higher, the better)
- a plastic cup
- a spoon or pestle to crush the cereal (better still, a blender)
- hot water
- a very strong magnet
- clean white paper
- a clear, sealable plastic sandwich bag

What do I do? Fill the cup to about two-thirds full with cereal, and with the spoon or pestle crush the cereal into a fine powder. It is worth spending a lot of time on this stage – the finer the powder, the better. Put the crushed cereal into the sandwich bag and add hot water. Leave the mixture for about 15 to 20 minutes. Now gently tilt the bag forward so that the cereal collects on one side, and place the magnet along the outside of the bag near the cereal, running it over the bottom, because the iron tends to sink. Tilt the bag back so that the cereal runs away from the magnet. You can also lay the bag flat on the table and stroke it with the magnet towards one corner.

Alternatively, if you are using a blender, put the cereal straight into the blender and add hot water until the cereal is submerged. Wait for about 15 to 20 minutes until the cereal is soft, then blend it all together. While the blender is whirring, place the magnet on the outside of the blender near the mixture, and keep it there as you turn the blender off.

What will I see? The magnet will attract a black fuzz of iron. Move the magnet over the surface of the bag or blender and the tiny pieces of iron will follow it.

What's going on? The black stuff really is iron in your cereal – the same stuff that is found in nails and trains and motorbikes. And it's quite heavy, which is why you need to make sure you run your magnet along the bottom of the cup. The iron is added to the mix when the cereals are being made and you really do eat it when you devour your cornflakes.

The reason it is added in a form that you can extract is that iron ions (iron that would more easily combine with other molecules in the cereal) increase the spoilage rate of the food. Using iron in its pure metal form gives the cereal a longer shelf-life.

The hydrochloric acid and other chemicals in your stomach dissolve some of this iron and it is absorbed through your digestive tract, although much of it remains untouched and is excreted.

PS: Humans need iron for many important bodily functions. Red blood cells carry haemoglobin, of which iron is a key constituent. Haemoglobin transports oxygen through the blood to the rest of the body by binding oxygen to its iron atom and carrying it through the bloodstream. As our red blood cells are being replaced constantly, iron is an essential part of our diet.

Saucy stuff

What's the best way to get tomato ketchup out of a bottle without breaking the bottle or splattering the sauce?

Some would insist, of course, that getting ketchup out of a bottle is an art not a science. We disagree for, as you will discover, there seems to be a lot of physics involved. The arrival of squeezy bottles has diminished the fun and, for purists, the glass ketchup bottle remains the ultimate challenge…

What do I need?

- a meal requiring tomato ketchup (it's not essential, as you can do this experiment using an empty plate, but there's no doubt that chips enhance the experience)
- a glass bottle of tomato ketchup

What do I do? Attempt to shake the ketchup out of the bottle onto your plate without splattering it or breaking the bottle. Ideally, you'll be able to release the sauce in pleasing globs that form an ordered pile in one part of the plate.

What will I see? It depends on your skill. If you follow one of the methods described below, you may be successful in getting all the ketchup to land on your plate. Technique and practice seem to be the key.

What's going on? Science is, of course, involved in extracting the ketchup from the bottle. The worst thing you can do, as the first of our methods demonstrates, is to bang the bottom of the bottle to get the ketchup moving. While you will see this technique employed widely in fish and chip shops, it is of little use. Instead, you should turn yourself into a centrifuge or exploit the thixotropic nature of ketchup.

Here we present a few of the best methods supplied to us by *New Scientist*'s readers, each named after the writer. You should try them all before settling on the one that suits you best. All, as you would expect, employ sound scientific principles and all have their champions among the magazine's staff.

The Foy (or inertia) method: Most people hit the bottom of the upturned bottle, which only ensures that the inertia of the sauce sends it in the opposite direction – relative to the bottle – to the one you want it to go. The sauce is pushed back into the bottle, rather than out of it. Instead, to get the last drop of ketchup out, hold the upturned bottle over your plate and hit the underside of the wrist of the hand holding the bottle with your other fist, jerking the bottle upwards. The inertia of the sauce will now eject it from the bottle.

The Wong (or centrifuge) method: First, put the lid on the bottle and grip it at its base. Then swing your arm as if you were throwing a ball overarm. This method, which uses the principle behind a centrifuge, forces the ketchup to the top of the bottle, allowing you to pour it out. (Whether you can use such a flamboyant technique in a posh chippy is open to question.)

The Lloyd-Evans (or thixotropic) method: Ketchup is gloopy because it is thixotropic. This means that, when it is at rest, it has a thick gelatinous consistency that can be altered to a runny consistency by the input of energy, typically by shaking. (This is in contrast to dilatant materials – see the 'Mixing madness' experiment on page 161.) The thixotropy is provided by the starch used in ketchup. Starch molecules come in the form of long chains and, when starch powder is mixed with water and heated or subjected to enzyme treatment, weak links are formed between the long molecules. This is what happens when the ketchup is made at the factory and there is a physical change in the ketchup, creating a pasty, gelatinous matrix. These thickening and gelling abilities of starches such as corn flour, rice flour, potato flour and powdered arrowroot are used in sauces, gravies and soups, as we can see in the 'Shape shifting' experiment on page 62. To get the ketchup out of the bottle, first ensure that the lid is on, unless you want to upset the person sitting directly opposite you, then give the bottle some vigorous but not over-athletic shaking. This will break some of those weak bonds between the starch molecules. Now turn the bottle upside down over your plate and watch the ketchup emerge in a slow, gentle stream.

The Bellis (or vibration) method: Rapidly hit the side of the bottle with the soft edge of your fist while holding the bottle at an appropriate angle. The vibration of the rapid thumping will break down the structure as above, allowing the ketchup to flow easily. This is the same principle used to settle concrete into moulds. In the case of concrete, however, a vibrating device is probed into the mix and the rapid vibrations shake the concrete, allowing it to spread downwards into the mould.

The Hann (or temperature) method: To get a nice even coating on your chips the best thing you can do is to warm up the ketchup. Provided you remove the metal cap, a burst of *no more* than 15 seconds in a microwave usually does the trick. Obviously, the amount of sauce in the bottle and how chilled it was will affect the success of this method, but heat is a guaranteed way of making the ketchup runnier as the bonds between layers of molecules become easier to break with shear stress and slide past each other more easily as the temperature increases. This property of viscous fluids – the phenomenon by which their viscosity tends to fall (or, alternatively, their fluidity increase) as their temperature increases – is known as the temperature dependency of liquid viscosity.

The Goldstein (or poking) method: Take a long, thin object (a chopstick works well) and poke a hole into the ketchup through the mouth of the bottle. This breaks the air/liquid seal and lets the sauce flow freely.

The Medhurst (or 'what-not-to-do') method: Leave the bottle in the back of your store cupboard for a few years until fermentation sets in. Pressure will have built up inside the bottle so that when the lid is removed the sauce will eject dramatically from the opening. This certainly removes the ketchup from the bottle but whether it fulfils the 'no splattering' criterion is open to debate. Howard Medhurst, who suggested this method, admits that his kitchen now has a 10-cm-wide red streak across the entire ceiling, one wall and half of the floor.

PS: Ogden Nash, renowned for his witty light verse, had something to say on the subject. His granddaughter, Frances R. Smith, of Baltimore, Maryland, who is the family representative of Nash's back catalogue, kindly supplied us with the following poem: 'The Catsup Bottle' (catsup being an American variant spelling of ketchup):

First a little
Then a lottle
The catsup bottle.

Bouncing rice

Why does a grain of cooked rice in a glass of fizzy lemonade repeatedly rise to the surface and fall to the bottom?

The author has fond memories of fooling gullible schoolmates with this 'magic' trick, telling them that for a modest sum he could make objects at the bottom of the glass mystically rise to the surface. That was until he was outed by the head of chemistry and forced to hand back his ill-gotten proceeds.

What do I need?

- a glass of fizzy lemonade (other fizzy drinks will work just as well, but transparent drinks are better than opaque ones if you want to see the rice)
- grains of cooked rice (uncooked rice also works, but takes longer as it has fewer nucleation sites), or citrus fruit pips if you enjoy a slice of lemon in your drink

What do I do? Drop the grain of rice into the lemonade and wait. If you'd like to run a control alongside this experiment (for the importance of controls see 'Fizz fallacy' on page 28, put a grain of rice in a glass of still water.

What will I see? The rice will sink to the bottom of the glass. Then, after a short while, it will begin an inexorable cycle of rise and fall between the surface and the glass bottom. The control will do nothing.

What's going on? Initially, the grain of rice sinks because it is denser than the liquid around it. However, once it has reached the bottom of the glass, bubbles of carbon dioxide begin to collect around the grain. This is because bubbles in fizzy drinks tend to form preferentially around rough surfaces which provide nucleation sites for their formation (see 'Over the top' on page 7).

Eventually, the bubbles of carbon dioxide begin acting as if they were airbags inflated underwater and attached to the grain. The rice then becomes buoyant and starts rising. When it reaches the surface the bubbles burst and it sinks again, restarting the process. If you leave the lemonade and rice for long enough, the drink will go flat and the rice will stop ascending.

PS: We are indebted to the students of the Aberdeen University Scottish Country Dance Society who were kind enough to send in their extensive research in this field carried out

①
bubbles form on rice grain in lemonade

②
buoyant rice grain rises to surface

③
pop
bubbles pop at surface and rice grain sinks again to start the process once more

over a number of years in controlled laboratory conditions (the students' union bar) using a variety of drinks and bar snacks. They tell us that the most consistent effects were usually obtained using salted peanuts and cheap lager. In fact, a handful of peanuts rising and falling in a pint of lager can produce an effect similar to the lava lamps popular in the 1970s and described in the 'Oil lamp' experiment on page 109. Unfortunately, but not surprisingly, the amount of salt on the nuts makes the cheap lager even more unpalatable, so only do this if you have no intention of consuming the drink later.

Interestingly, Newcastle Brown Ale and many draught or bottled ciders are so fizzy that bubbles form too quickly and the salted peanut never has time to sink back to the bottom but remains at or near the surface, while dry-roasted peanuts are far too dusty to work properly and pork scratchings and crisps don't sink in the first place, but float soggily at the surface.

Tuneful cuppa

Why does the note coming from a spoon repeatedly hitting the inside of a newly poured cup of instant coffee or hot chocolate rise in pitch?

This is the kind of thing you have always been subconsciously aware of yet never seen put into words. But now you know about it you'll just have to try it. So make yourself a nice cuppa and get stirring ...

What do I need?

- a large ceramic mug
- instant coffee or powdered hot chocolate
- boiling water
- a teaspoon

- milk
- sugar

What do I do? Put the powdered coffee or hot chocolate into the mug, add the boiling water and stir thoroughly. Hold your spoon vertically in the liquid and listen carefully as you repeatedly tap the base of the mug for a minute or so.

What will I hear? The sound of the spoon hitting the bottom of the mug gradually rises in pitch.

What's going on? The reason why this happens has been a matter of debate for some time.

When the mug is tapped it acts somewhat like a bell: the sound generated is determined mainly by the material of the mug and its thickness, diameter and height. It is also determined by the distance between the surface of the water and the bottom of the mug, which causes the mug to resonate at a particular frequency. It is tempting, therefore, to speculate that stirring the water, which causes it to sink slightly in the middle and rise a bit along the sides of the mug, is responsible for the change in pitch. But this slight change in water level is not nearly enough to account for the big change in pitch, which can sometimes exceed a full octave. Also, the pitch change occurs even when the water is not rotating, so something else must be going on.

And that something is air. When you pour hot water on instant coffee, many of the volatile components in the coffee form tiny bubbles. Additionally, water frequently has air dissolved in it, so when it is added to powders such as coffee, nucleation sites form around the grains and more bubbles are created (see the 'Over the top' experiment, page 7). The air bubbles lower the speed of sound in the water by temporarily making it more compressible (less 'springy'). It takes longer for the sound wave to travel between the bottom and the surface, and thus the pitch falls when the instant coffee is first added. Then, as the bubbles rise to the surface and burst, the sound travels faster and the pitch rises.

The greatest difference in pitch occurs when the spoon is tapping the bottom of the mug. This is because the sound waves have further to travel from the base of the mug to the surface of the liquid than they do from the sides.

PS: Because seemingly thousands of cups of coffee or tea are consumed in the *New Scientist* office each week, staff were asked to experiment with their daily cuppas. They found the effect in cups of tea was tiny however much milk was added or however vigorous the stirring. But in instant coffee or powdered hot chocolate there was a clear result.

Over a period of about 5 seconds after the boiling water was added to the powder, the note of the spoon against the mug rose distinctly as the drinks were stirred. Bubbles could be seen rising to the surface, which suggests that they were the key factor in the change of sound. Tea produces fewer bubbles than instant coffee or hot chocolate, which explains why the effect was more difficult to reproduce in tea.

Want to read more? Frank F. Crawford's paper 'The hot chocolate effect' was published in the May 1982 edition of the *American Journal of Physics* (vol. 50, p. 398).

Green eggs and cabbage

Why does the juice from cooked red cabbage turn fried egg white green?

Chef Heston Blumenthal, one of the founding fathers of molecular gastronomy, would love this one. So would Dr Seuss. You only need a small amount of red cabbage juice to change the colour of fried eggs from white to green. Children and guests will be completely taken aback.

What do I need?

- shredded red cabbage (boiled for 20 minutes)
- a frying pan
- oil
- one egg

What do I do? Squeeze the juice from the cooled cooked cabbage into a jug. Heat the oil in a frying pan, begin to fry the egg until the white is just turning from clear to white. Drip a small amount of cabbage juice into the setting egg white.

What will I see? The egg white will turn lurid green where the juice hits it.

What's going on? Red cabbage juice is a good indicator of whether a substance is an alkali or an acid. If added to an alkali, such as ammonia, it will turn green; if added to an acid, such as lemon juice, it will turn red. In neutral substances it is purple, the natural colour of red cabbage. Because egg white (mostly the protein albumen) is alkaline, it turns green. Any number of substances can be tested in this way, although take care to avoid strongly corrosive chemicals such as drain cleaner or bleach because these can be dangerous.

The experiment works because red cabbage contains water-soluble pigments called anthocyanins (also found in plums, apple skins and grapes). These change colour depending on whether they are in the presence of an acid or an alkali. These change the number of hydrogen ions attached to the molecule – acids donate hydrogen ions while alkalis remove them – and it is the presence or absence of hydrogen ions that is responsible for the different colours. This explains why red cabbage that is pickled turns red, rather than its natural purple colour. Pickling takes place in vinegar, which is acidic.

Red cabbage juice breaks down quite quickly, so if you are going to use it to test the acidity or alkalinity of other household foods or products use it sparingly and fast.

PS: One trick that will prove popular with children is to create 'magic' paper. Soak cheap, absorbent paper in boiled red cabbage water and leave it to dry. Then paint it with household substances such as vinegar, orange juice or washing powder dissolved in water. A range of colours will appear depending on the acidic or alkaline nature of the 'paints'.

Burnt offerings

Why does brown or whole wheat bread char more quickly than white bread when you toast it? (If it does, it explains why the author burns the toast but his wife does not.)

What do I need?

- slices of brown or whole wheat bread
- slices of white bread
- a toaster
- butter plus honey or marmalade to ensure your efforts are not wasted

What do I do? Place the brown bread in the toaster. Check it every 15 seconds to see when it starts to burn (you'll need to introduce an objective measure of what you consider to be charred bread). Take a note of the time. Repeat with the white bread.

What will I see? You'll notice that the brown or whole wheat bread toasts much faster than the white bread.

What's going on? During heating, a complex reaction occurs between the proteins and sugar contained in the bread. This is known as the Maillard reaction, and it produces the typical flavour we know from toasted bread as well as the colour formed as the bread toasts.

The Maillard reaction is a chemical one between an amino acid and a reducing sugar. It usually requires the addition of heat, as in the case of toasting bread. The reaction is widely used in the food industry for flavouring, producing different flavours and odours according to the type of amino acid involved in the reaction.

Because brown bread and whole wheat bread contain more sugar and protein than white bread, they undergo more rapid Maillard browning than white bread.

One other factor may help to explain why white bread chars more slowly, and this is albedo – the proportion of incident light or radiation that is reflected by a surface. White bread reflects more radiation than brown bread, which is why it appears whiter. Because darker breads absorb more radiation in the form of heat from the toaster, they heat up more quickly and therefore burn quicker.

PS: While you have the loaves of bread out and the jar of honey ready, you can try another experiment. Take a fresh, untoasted slice of bread and spread it with honey. Then set it aside. If you leave it for a few minutes, you'll see that it becomes concave on the side spread with honey. This is because bread is approximately 40 per cent water, while honey is a strong solution containing approximately 80 per cent sugar. Sugar is hygroscopic, which means that it soaks up moisture. This causes moisture to be drawn out of the bread and into the honey by osmosis. Extracting the water makes the bread shrink, but only on the side exposed to the honey. This explains why the bread becomes concave.

This is less likely to happen if you butter your bread before spreading the honey. Butter forms a fat-rich, water-impermeable layer that protects the bread from dehydration by the honey.

Want to read more? The chemistry of different bread types is discussed in *The Composition of Foods* by Robert Alexander McCance and Elsie Widdowson (Elsevier/North-Holland Biomedical Press).

Silver lining

If you hold an egg in a candle flame until it becomes sooty and then dunk it in water it looks silver. Why?

This is an old experiment but an extraordinary one – and it allows you to use a hopeless pun as you point out that 'every egg has a silver lining'. What's more, because it's another question involving eggs you can try it out alongside the 'Green eggs and cabbage' and 'Sucking eggs?' experiments on pages 47 and 102 respectively.

What do I need?
- a white or pale brown hen's egg
- matches
- a candle
- water
- a glass bowl

What do I do? Light the candle and hold the egg between your index finger and thumb. Because you are going to move the egg into the flame this is an experiment children will have to watch from the sidelines. Hold the egg near the top of the candle flame so that it begins to become covered in soot. Rotate it so that as much of the egg as possible becomes sooty. Fill the bowl with enough water to cover the egg and carefully place the egg in the water.

What will I see? It's quite astounding. Once in water the egg looks neither black nor egg-coloured, but silver with an almost mirror-like quality.

What's going on? The silver appearance of the egg's surface is created by a very thin layer of air that gets trapped between the egg and the body of the fluid, because the soot-covered surface repels liquid. The air forms a film around the egg. Light passing through the water strikes this layer of bubbles and is reflected back, like a mirror. Eventually, the air bubbles will dissolve into the water and the shell will appear sooty again.

PS: The good news is that if you use the eggs immediately after carrying out this experiment they are still edible. If you wanted to be truly ecologically sound you could even use the water from the bowl to boil them in.

Strings attached

Most hard cheeses such as Cheddar go stringy when grilled, yet halloumi doesn't melt at all, it just chars and maintains its shape. Why is this?

All cheese is made from milk, so there's no obvious reason. When you try this out, keep a slice of toast from the 'Burnt offerings' experiment (page 49) to hand – there's no point in wasting melted cheese…

What do I need?

- cubes of Cheddar
- cubes of halloumi
- kebab skewers
- a grill
- bread to eat the results with

What do I do? Thread the cheese cubes onto the skewers, place under the grill, heat at a high temperature and watch what happens.

What will I see? The Cheddar will start to go stringy and drip into the grill pan. This is where your bread, slowly toasting below the cheese at the bottom of the grill pan, will come in handy. Meanwhile, the halloumi will retain its cube shapes. Fortunately, this too can be eaten after it has charred and cooled a little.

What's going on? Uncooked Cheddar contains long-chain protein molecules curled up in a fatty, watery mess. It also has a low melting point, which means it becomes runny before it burns and as the fats and proteins melt and drop under the weight of gravity, they are dragged into dripping strings – just as they are if you take a bite of the cheese after it has settled onto your toast. The long-chain protein molecules are unravelling and forming fibres as the cheese softens. If you are a fan of pizza you'll know that mozzarella works even better and can string out to 30 cm or more. In fact, string length is a measure of protein content – the longer the string of cheese, the more protein it contains.

However, in addition to melting and becoming stringy, the protein in Cheddar is denatured by the heat. This is why previously molten Cheddar turns into an unpleasant rubbery lump after it cools. The makers of halloumi and similar cheeses such as paneer take advantage of this process. Halloumi has already been heated and partially cooked during production, so important changes have already taken place before you put it under your grill. The denaturing effect means it is already in a rubbery state and so keeps its shape, making it perfect for use in kebabs.

Of course, it can still burn easily, so keep a close eye on this experiment at all times. And, although it will ruin the experiment, any fool knows that if you grate your Cheddar and add a beaten egg, half a teaspoon of mustard powder and a pinch of ginger, the whole thing will taste 100 times better anyway. After you've got bored with experimenting, add these for a satisfying lunch.

PS: Try reheating cold, previously molten Cheddar to see whether its properties have become like those of halloumi. You'll find you've produced your own denatured cheese in the kitchen, even though it's just about inedible.

In the dumps

Why do cooked dumplings float while uncooked ones remain close to the bottom of the pot?

We have happy childhood memories of home-cooked dumplings, and this question offers us the perfect opportunity to revisit a hearty dish of beef and beer stew with a few delicious dumplings floating on the top. The art of dumpling making seems to have declined in recent years, probably due to contemporary healthy eating fads, so if you are willing to have a go, we've provided a recipe below courtesy of the author's mum, who has always been a dab hand at traditional fare.

What do I need?

- uncooked dumplings
- a pot of simmering liquid (preferably beef and beer stew or a decent vegetable soup, but you could get by with a pan of simmering water if you don't mind missing out on a sublime culinary experience)
- a slotted spoon

What do I do? Lower the uncooked dumplings carefully into the simmering liquid.

What will I see? The uncooked dumplings will initially sit on the bottom of the pan. As they cook over the course of the next 20–30 minutes they will rise to the surface. While we admit that this is easier to see if you are using boiling water, we still think that it's more civilised to cook the dumplings in the stew or soup, and of course, you'll end up with dinner after the experiment has been completed.

What's going on? Two factors are at work. It seems the main reason the dumplings float once they are cooked is because bubbles become trapped in the dough matrix of the dumplings – held together with the gluten in the flour – and provide buoyancy. Similar foodstuffs, such as meatballs made from minced meat, egg, breadcrumbs and herbs, or matzo balls made from eggs, chicken fat, water and matzo meal, have the same propensity. First they sink, then they rise. This is because the air spaces bound up in their matrix expand when they are placed in boiling liquid, making the mixture less dense. This, of course, makes them buoyant. Consequently, if left to cool, both meatballs and dumplings will return to the bottom of the pan as the air spaces shrink, so the golden rule is eat them while they are hot, they taste far better that way.

There is, however, another factor at work, especially in the case of dumplings, and that is the presence of baking powder. Dumplings are made with plain flour and a large quantity of baking powder, or with self-raising flour. Baking powder is a leavening agent containing an alkali, most commonly sodium bicarbonate, and an acid, plus starch to stop the mixture becoming damp. When it is dissolved in water the alkali reacts with the acid to produce carbon dioxide, which creates bubbles in the dumplings. Self-raising flour has a similar effect, although in this case the leavening agent is already part of the mixture and there is no need to add baking powder.

Try making dumplings without a leavening agent and you'll find they are far more reluctant to float, although, given enough time and heat, the air spaces in their matrix will ensure they eventually reach the surface.

PS: This recipe has been in the O'Hare family for generations. We cannot claim its superiority over other dumpling recipes, but we do know that it leads to dumplings that follow the experimental rules outlined above. First they sink, then they float...

150 g self-raising flour
75 g shredded suet (you can use a vegetarian version if you prefer)
decent-sized pinches of salt and pepper
1 teaspoon dried herbs
approximately 3 tablespoons of water

Sieve the flour into a bowl and mix it well with the suet, pepper, salt and herbs. Add the water a little at a time to give a slightly soft but not gooey dough. Lightly flour your hands and form your dough mixture into 6–8 balls. Carefully place them in the simmering stew, soup or water using a slotted spoon and allow them to cook for 25–30 minutes. Check them after 20 minutes – if the cooking liquid is boiling vigorously, they'll be cooked sooner.

If you have recipes of your own, you can compare and contrast the buoyancy or otherwise of your own dumplings with our version.

Teething trouble

Why does eating spinach make your teeth feel weird?

Sadly, this works better with canned spinach than with fresh spinach, but it is still one way to get the kids to eat green vegetables. However, be cunning: make sure that they think they are doing the experiment rather than being experimented on…

What do I need?

- plenty of spinach leaves (preferably canned, heat-processed spinach)
- a pan of boiling water
- a colander
- a plate
- a fork

What do I do? Boil the spinach until it is cooked, drain it, let it cool a little and eat it. Then run your tongue around your teeth and mouth.

What will I feel? Your teeth and the inside of your mouth will feel fuzzy and furry. This may be one of the reasons why children particularly dislike spinach, but once they realise they are part of a noble experiment their attitude changes radically.

What's going on? Spinach contains a large amount of oxalate crystals – mineral salts of oxalic acid. When spinach is cooked – especially the canned variety – some of the spinach cell wall structure is damaged and the oxalate crystals leak out. These coat your mouth to give the fuzzy feeling, and it explains why fresh, uncooked spinach does not produce a similar effect. Spinach is also rich in calcium and oxalic acid, and these combine with the calcium in saliva to deposit large amounts of furry, calcium-rich plaque on your teeth.

PS: Chard and beetroot leaves have a similar effect.

Warning: People who have problems with renal function should avoid spinach and other oxalate-rich foods because of the increased risk of producing oxalate stones in their kidneys.

Floaters

Why do floating pieces of breakfast cereal stick to each other and to the sides of the bowl?

While it has been suggested that Rice Krispies and Cheerios herd together as a defensive strategy to avoid being eaten, this clearly is not successful. Other forces are at work…

What do I need?
- a bowl of milk
- a wheat- or rice-based breakfast cereal
- tap water
- three polystyrene cups

What do I do? First check out the basis of the question. Pour a bowl of milk and add the breakfast cereal of your choice. Don't overfill the bowl; just add enough to cover a third of the surface at most. Then work out the physics involved using the polystyrene cups.

Fill the first cup to within 1 cm of the rim. Fill the second cup to the top and then carefully add more until the meniscus is above the top of the cup but not spilling over – in other words, the water is held in a convex bulge above the top of the cup by surface tension.

Now place two 1 cm diameter circles of polystyrene torn from the third cup in the middle of each of the first two.

What will I see? In the first experiment the breakfast cereal will clump together, preferring the edges of the bowl to the centre. You can now eat the cereal.

In the second experiment the piece of polystyrene floating in the partially filled cup will, with a little prompting, move to the side of the cup and be held there. By contrast, the piece floating on the convex bulge of the water in the second cup will remain near the centre. Furthermore, if you push the piece to the edge of the cup with the tip of a pencil, the edge repels the small piece towards the centre with considerable force.

What's going on? This is caused by the surface tension of the water. In the partially filled cup the water surface curves up to meet the polystyrene wall because water molecules are more attracted to polystyrene than to each other.

The water forms the convex bulge at the top of the full second cup because the surface tension constrains the liquid surface to the smallest area possible, which similarly accounts for the spherical shape of liquid drops. The water also curves up to meet the small circle of polystyrene on all sides. Where the water meets the polystyrene of the small circle, the surface tension pulls on each contact point in a downwards and outwards direction determined by the angle of contact with the water. When the circle is in the middle of the cup the pull on the circle on one side is directly balanced by the pull on the opposite side, because the water curves up to meet the circle equally at all points. This means the polystyrene circle is held in the middle of the liquid in the cup.

However, when the piece is moved towards the side of the partially filled cup, the upward curve of the water surface at the side of the cup reduces the curve of the surface in contact with the circle as these two curves come together. This increases the outward pull on the side of the circle nearest to the cup edge, resulting in a net force towards the side of the cup. As a result, if the cup is only partially full, the polystyrene circle will move away from its central position and sit at the side, unlike the polystyrene circle in the full cup.

The effect accounts for the clumping of those cereal pieces on the surface of milk in your bowl and also for the similar behaviour of leaves and twigs on ponds and lakes.

Apple fool

Is it true that you can fool yourself into believing you are eating something other than the food that's in your mouth?

It's certainly true that the senses can be fooled. The author has personal experience of how one sense can be perceived as another. When a virus temporarily took away my sense of smell I was astounded how easy it was to confuse one sensation with another. I'd often leave an air-conditioned office on a freezing day, sense the chilly air, and think, 'Wow! My sense of smell has returned.' Somehow, my brain was confusing the sense of touch which detected the cold with the return of my sense of smell. This experiment shows how easily human sensations can be fooled. And, after fooling yourself, you can read about synaesthesia…

What do I need?

- a sliced apple
- a sliced pear
- your nose
- your mouth

What do I do? Hold a piece of the pear under your nose while eating a slice of the apple.

What will I taste? Despite the fact that you are eating apple, you'll think you are eating pear.

What's going on? The confusion arises in part because most people think that they taste using their taste buds. They don't. We all detect flavours via our sense of smell.

Indeed, most of what we call taste is actually flavour produced by the smell of food passing from our mouths into our nasal cavities where we detect it through our sense of smell, or olfaction. True taste is only the bitter, sweet, salt, sour and umami (savoury) detected by the taste buds. Bite into a strawberry and your tongue only tells you that it is sweet, just as it would if it were chocolate. It's the odours rising through your throat to your nose that tell you that that particular sweetness is strawberry-flavoured.

If you remove the sense of smell or block those odours, or – as in the experiment above – replace them with something else, you can confuse your senses of taste and smell. Your taste buds detect the apple in your mouth as being sweetly acidic using its sweet- and sour-detecting taste buds, and your sense of touch recognises the texture to be that of a type of tree fruit. However, because you can more strongly detect the smell of pear under your nose than you can the flavour of apple rising through the back of your mouth, your brain is fooled into thinking that what's in your mouth is a piece of pear.

The experiment works best this way round because pears have a stronger smell than apples.

PS: People who have synaesthesia make bizarre sense associations all the time. No one is really certain, but this condition is thought to involve crossed wires in the brain. We are born with many connections to different areas of our brain, but as we mature our brain cuts back on those it doesn't need. Synaesthetic connections are possibly some of those that were not pruned out during infancy. The extraordinary result is that some people – as many as 1 in 25 – perceive things like letters, words, numbers or days of the week as having very distinct and immutable colours. A smaller group even experience sounds as having colour or tastes as having shapes. One member of this group was reported as saying that she knew when the milk was about to go off because its shape began to quiver.

Shape shifting

Why is corn flour mixed with cooking oil attracted to a balloon?

Corn flour, as you will also see in the 'Mixing madness' experiment on page 161, has many uses in scientific demonstrations, with its wonderful starchy properties giving rise to a number of weird effects. This experiment is also related to the effect in the 'Wayward water' experiment on page 127 and shows the strength of even very small electrical charges.

What do I need?

- corn flour
- cooking oil
- a wooden spoon (the spoon must be wooden because you need an electrical insulator)
- an inflated balloon

What do I do? Mix the oil and corn flour together until the mixture resembles a thick cream. Rub the balloon on your clothing to create a charge on its surface as in the old party trick of rubbing a balloon and sticking it to the wall using the static electricity generated by the friction. Pick up some of the corn flour/oil paste on the wooden spoon. This works best if a few stable dribbles are hanging from the spoon. Slowly move the spoon and its dribbles closer to the balloon.

What will I see? When the spoon gets close enough the paste thickens and moves, independently of the spoon, towards the balloon.

What's going on? When you rub the balloon, the electrical charge you create on its surface is positive. When the tiny particles of starch that make up the corn flour come near this positive charge, their negatively charged electrons are pulled towards the balloon – electrical charges act like magnetic poles, with a negative charge attracting a positive one and vice versa. The sides of the starch particles facing the balloon are now more negative (and consequently the other sides of the starch particles facing away from the balloon are more positive). The paste moves towards the balloon because of the attraction of the positive balloon charge and the negative starch charge on the balloon-facing sides of the starch particles. The positive sides of the starch particles have no balloon or similar charge to hang on to and are therefore unable to stop being dragged along for the ride.

The whole mixture thickens because, although the charges cannot move between starch particles as they are insulated by the oil, they can all pack up next to each other. The negative sides of the particles nearest to the balloon are attracted to the positive sides facing away from the balloon and therefore all the particles pack more tightly.

PS: Corn flour is an incredibly versatile product. Produced from maize, it is starch-rich, hence its name in the USA: cornstarch. It is widely used for thickening sauces, but its great culinary benefit is that it only works when heat is applied. If you simply add it to a small amount of cold water no thickening occurs. However, pour this mixture into a heated soup or casserole and the water and starch molecules begin to bond. The starch particles enlarge, trapping the water as they grow. At about 65 °C the structure of the starch breaks up and forms a mesh of bonded starch and water molecules, preventing free movement of the water molecules and creating a thick sauce.

Aim and pour

Why does milk dribble down the underside of its carton if you pour it too slowly?

It's not only milk, of course. Many liquids poured from cartons, such as orange juice and soup, result in a sticky floor or ruined shoes. And it's difficult to avoid the effect because when the carton is full you have little choice but to tip it gently when trying to fill your glass.

What do I need?

- cartons of milk or other liquids
- glasses
- a cloth (for cleaning up afterwards)

If you want to demonstrate the effect further you'll need:

- a vertical cylinder (a washing-up liquid bottle or a wine bottle will do)
- a lighted candle

And if you are feeling very confident you'll need:

- a hairdryer
- a table tennis ball

What do I do? Open the carton and pour the milk into the glasses. Vary the speed at which you pour so that you range from just a dribble to a rapid stream.

What will I see? At low pouring speeds the milk will cling to the edge of the carton and dribble its way down the container before depositing liquid on the floor – this is where the cloth will come in handy. At faster speeds the liquid will pour freely, allowing you to fill your glass with aplomb.

What's going on? When the carton of liquid is tipped, the surface of the liquid in the container is raised and moves towards the opening of the container. As the carton is tipped further, liquid pours from the opening, creating pressure at the opening. In addition to this pressure force, there are surface tension forces acting on the fluid that tend to draw it towards the surfaces of the container. At high pouring speeds, the pressure force is much greater than the surface tension forces, and the fluid will leave the carton in an orderly fashion, following a predictable parabolic path towards your glass.

However, at low pouring speeds, a point is reached where the surface tension forces are sufficient to divert the path of the fluid jet so that it fails to leave the opening cleanly and becomes attached to the top face of the carton. Once attached, a jet of liquid will tend to stick to that surface thanks to the surface tension forces and a phenomenon known as the Coanda effect. This occurs when a fluid jet on a convex surface (such as a water jet from a tap curving around the back of a spoon) generates internal pressure forces that effectively suck the jet towards the surface.

The combined result of surface tension and the Coanda effect enable an errant flow of fluid to negotiate the bend from the top face of the carton onto the carton's side and, ultimately and rapidly, onto the floor – or your shoes.

The Coanda effect (also known as 'wall attachment') is named after Henri Coanda (1886–1972), who invented a jet aircraft that was to be propelled by two combustion chambers, one on either side of the fuselage and pointing backwards. These were positioned near the front of the aircraft. To Coanda's horror, when the engines were ignited, the jets of flame, instead of remaining straight and pointing directly out of the back of the engines, clung to the sides of the fuselage all the way to the tail. While obviously troublesome at the time, Coanda has been immortalised by this discovery with the effect now bearing his name.

We must also take into consideration one more effect that causes erratic emission of liquid from cartons. This is the 'glugging' that occurs as air is sucked into the carton to replace the lost fluid. This causes the fluid jet to oscillate, leading to intermittent surface attachment and more wet shoes even at relatively high pouring speeds.

PS: The Coanda effect is seen in many circumstances because of the general tendency for fluid flows to wrap around surfaces. Another demonstration that you can try consists of taking a vertical cylinder (this is where the washing-up liquid bottle or wine bottle come in handy) and placing a lighted candle on the far side. When you blow against the near side of the bottle the candle is extinguished despite being apparently protected from the breeze, because the current of air wraps around the bottle and links up again on the far side rather than being deflected away.

Now take the hairdryer, switch it on to a cool setting and aim the air stream vertically upwards. The effect works best if the hairdryer's nozzle is roughly the same size as the table tennis ball. You'll be able to take the table tennis ball and position it in the air flow where it will bob about quite happily without falling off (you may need a few attempts to position it because finding the right spot can be difficult). Again, this is a case of the air stream sticking to a surface, this time to the table tennis ball, and the ball is held in place by the Coanda effect. Because this effect is so powerful, you'll have to tilt the hairdryer steeply away from the vertical before gravity can win out.

Want to read more? A photograph of the Coanda, the first true jet aircraft, built in 1910, can be found at www.allstar.fiu.edu/aero/coanda.htm alongside information about the designer.

Stirring stuff

If you are stirring a drink such as leaf tea, why do the leaves head to the middle of the teacup the faster you stir?

Intuition suggests that objects should fly outwards not inwards, so what's going on? There's no truth in the rumour that tea leaves are possessed of supernatural qualities as some fairground entertainers would have you believe, so we have to fall back on that old stalwart of the truth – real science.

What do I need?

- loose-leaf tea
- a cup or mug
- hot water
- a spoon

What do I do? Place the leaves in the mug, add water, stir the drink and watch what happens. If you feel thirsty you can always drink the tea afterwards (adding lemon, sugar or milk if you prefer), but if you are used to drinking only the product of teabags, be prepared to strain the leaves through your teeth.

What will I see? Any floating tea leaves will move to the centre of the mug as you begin to stir. As you increase the speed of stirring, they'll move more rapidly to the centre.

What's going on? The answer lies in what is known as the pressure-momentum balance in the rotating fluid. If the rotating fluid created by your stirring is to remain in the mug rather than breaking down the sides of the vessel, the inertial force created by the fluid rotation must be balanced by a pressure gradient within the liquid. Pressure is lowest in the centre and increases towards the wall of the mug. Think of a

As the mug is stirred, liquid moves outward across the top surface, down the outer walls of the mug to the bottom and up the centre of the mug to the surface again.

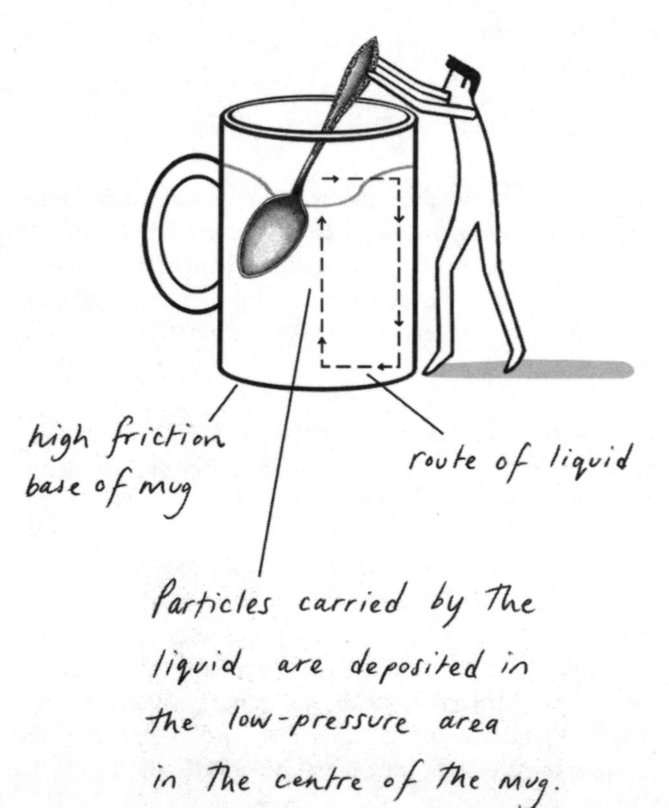

weight being swung around at the end of a piece of string: the tension in the string prevents the weight from flying off. This tension is the equivalent of the pressure gradient in the tea.

When you stir the liquid it is pushed outwards, raising the pressure at the edge of the mug and lowering it at the centre. You can clearly see that as the liquid rotates it is forced outwards, creating a dimple in the centre of the liquid surface. In fact, if it wasn't for the sides of your mug constraining the liquid, it would simply flow out over the tabletop. So any particle that is less dense than the surrounding liquid will tend to move inwards towards the centre of the mug where the pressure is lowest.

This makes sense, but even objects that are heavy move inwards. Try dropping a handful of peanuts into the liquid, or even small stones (don't use your best china for this experiment) and you'll see that they still cluster towards the middle when you stir the liquid.

The rough bottom of the mug induces more friction than other areas where the liquid is in contact with the walls, because the liquid is rotating parallel to this surface. Meanwhile, friction is least between the air and the rotating liquid at the surface, allowing it to move faster. So there's also a gradient between higher pressure at the surface and lower pressure at the bottom. This leads to a circulation current being formed – liquid moves out across the top surface thanks to the stirring motion, downwards nearer the outer wall, inwards along the bottom and back up the centre of the mug. Granules such as tea leaves or small stones are carried along with this current until they come to rest at the centre of the bottom of the mug where the pressure is at its lowest. Gravity also plays a part – while the liquid cannot accumulate in a pile at the bottom of the mug and so keeps on moving, flowing up the axis of rotation away from the bottom of the mug, the heavy particles are held at the bottom by a combination of the low pressure and the gravitational pull of the Earth.

Physicists call this accumulation of particles a boundary layer effect. The rough bottom of the mug is the key to inducing this property and if you can get hold of smoother vessels with lower friction than the average kitchen mug you'll see the effect is far less marked.

PS: The forces at work in your mug are the same ones that cause rubbish to collect in street whirlwinds on blustery days rather than being distributed evenly over the road.

Plastic milk

Can you make plastic out of milk and vinegar?

You would imagine that you'd need some pretty noxious, smelly chemicals to make plastic, but you can actually find the things you need to make malleable, doughy pieces of material in your own home. Instead of putting vinegar on your fish and chips and wasting your milk in your tea, use the two liquids to become a polymer chemist…

What do I need?

- a pint of milk
- a saucepan
- a sieve
- a spoon for stirring
- 20 ml of white vinegar
- rubber gloves
- water

What do I do? Pour the milk into the pan and gently warm it. When the milk is simmering (don't let it boil) stir in the white vinegar until you notice whitish-yellow rubbery lumps beginning to curdle in the mixture at the same time as the liquid clears. Turn off the heat and let the pan cool.

What will I see? First of all you'll smell the vinegary reaction, which is the key to this process at work. As the vinegar is added and stirred, the liquid gets clearer and the yellowy rubbery lumps form. When the pan has cooled you can sieve the lumps from the liquid, tipping the liquid down the sink. Put on the rubber gloves and wash the lumps in water. You can then press them together into one big blob – they will be squishy and will feel as if they are going to fall apart, but they will stick together after some firm kneading. You can now use your artistic skills to fashion the material into the shapes of your choice – *New Scientist* staff came up with balls, stars, a heart-shape for a pendant and even dinosaur footprints. Leave the material to dry for a day or two and it will be hard and plastic enough to paint and varnish.

What's going on? You have used the combination of an acid – in this case vinegar, which contains acetic acid – and heat to precipitate casein (a protein) from the milk. Casein is not soluble in an acid environment and so, when the vinegar is added, it appears in the form of globular plastic-like lumps. Casein behaves like the plastics that we see in so many objects around us, such as computer keyboards or phones, because it has a similar molecular form. The plastics in everyday objects are based on long-chain molecules called polymers. These are of high molecular weight and get their strength from the way their billions of interwoven criss-crossing molecules tangle together.

PS: Some forms of cheese-making rely on a similar technique – the name casein comes from *caseus*, the Latin for cheese. The Indian cheese known as paneer is made in a very similar way to the plastic you have just made, although in this case lemon juice is the acid used rather than vinegar. Afterwards, unlike our plastic milk, it is not dried out and allowed to harden to tooth-breaking consistency, and so remains soft and edible.

Sink or swim

Sauce sachets in a plastic bottle filled with water can be made to act like submarines. How?

We read about this and were astounded, so we tried it out in the office and it's true. You can make a sauce sachet – of the kind you find in pubs or fast-food restaurants – rise and fall in a plastic bottle just by squeezing the bottle. Why, and how? Science again provides the answer…

What do I need?

- a 2-litre plastic bottle
- water
- unopened sachets of ketchup, mayonnaise or any sauce

What do I do? Fill the bottle to the very top with water and add the sauce sachet. It helps if you have more than one sachet to experiment with because you need to find one that floats beneath – but just touching – the surface. Screw the bottle top on as tightly as you can. Then squeeze the bottle hard.

What will I see? The sauce sachet will sink to the bottom of the bottle. It's quite amazing. And when you release the pressure on the bottle it will rise again.

What's going on? When the bottle is squeezed, pressure is applied to the water. But, because liquids are generally very resistant to compression, this pressure is transferred to the sachet which, in addition to the liquid sauce, contains a small amount of gas. Gas is easily compressed, and as the sachet is squeezed by the surrounding water the space taken up by the sachet is reduced. As its volume shrinks, it becomes denser and reaches a point where it can no longer float. The sachet then sinks to the bottom of the bottle, returning to the surface only when the pressure falls after you stop squeezing the bottle.

PS: As a floating object falls or rises in liquid the forces around it change, as does the object's volume. At the surface it is less compressed (and therefore less dense) than it is at depth. In fact, a suitably compressible object can sit in a liquid at equilibrium, neither rising nor falling. Much of this was discovered by Archimedes in the 3rd century BC.

Try squeezing the bottle with differing pressures until you get the sachet to hover halfway between the top and the bottom of the bottle. (If you find this tiring, try using a G-clamp or even squeezing the bottle gently between a door and its jamb.) This is essentially how a submarine operates. To control its buoyancy a submarine has ballast tanks that can be filled with air or water. On the surface, the tanks are filled with air and the submarine's density is less than that of the surrounding ocean. To dive below the surface, the submarine's tanks are filled with water and the air vented until its overall density is greater than the surrounding water. A supply of compressed air (and therefore much reduced in volume from normal atmospheric pressure, similar to that in the sauce sachets after the bottle is squeezed) is kept on board and this is used for refilling the ballast tanks to allow the submarine to resurface as water is forced from the tanks. In order for the submarine to operate between the surface and the seabed, special trim tanks carry a fine balance of air and water. These carefully adjust the overall density of the submarine to allow it to achieve neutral buoyancy at whatever depth the captain chooses.

Citric secret

Why does lemon juice stop cut apples and pears from browning?

It might not improve their flavour, but it certainly works. And if you still want to eat the apples or pears afterwards, you can always wash off the lemon juice.

What do I need?

- apples
- pears
- a knife
- lemon juice
- two plates

If you want to take this experiment further you could also get hold of:

- some celeriac
- filter paper (you can cut up a paper coffee filter)
- apple juice
- vitamin C drink

What do I do? Slice the apples or pears so they have large areas of their inner flesh exposed to the air. Place half of them on one plate and half on the other. Sprinkle the fruit on one of the plates with lemon juice, and leave the other untouched.

What will I see? The fruit on the plate that has not been sprinkled with lemon juice will turn brown on its exposed surfaces much faster than the fruit covered in the juice.

What's going on? To understand why this happens we first need to understand why some plant tissues go brown in the first place. Plant cells have various compartments, including ones known as vacuoles and plastids, which are separated from each other by membranes. The vacuoles contain phenolic compounds which are sometimes coloured but usually colourless, while other compartments of the cell house enzymes called phenol oxidases.

In a healthy plant cell, membranes separate the phenolics and the oxidases. However, when the cell is damaged – by cutting, for example – phenolics can leak from the vacuoles through the punctured membrane and come into contact with the oxidases. In the presence of oxygen from the surrounding air these enzymes oxidise the phenolics to produce products such as polyphenol oxidase, which help to protect the plant and heal the wound – a process also mentioned in 'Banana armour' (see page 34). The drawback is that they also turn the plant material brown.

However, this browning reaction can be blocked by one of two agents, both of which are present in lemon juice. The first is vitamin C, a biological antioxidant that gets oxidised instead of the apple's phenolics. The second agent is organic acids, especially citric acid present in lemons. These make the pH lower than the oxidases' optimum level and thus slow the browning.

Lemon juice not only has more than 50 times the vitamin C content of apples and pears, it is also much more acidic than apple or pear juice, as a quick taste will tell you. This means lemon juice will immediately prevent browning.

You can prevent cut apples from browning without lemon juice by putting them in an atmosphere of nitrogen or carbon dioxide, thereby excluding the oxygen required by the oxidases, but this option is not open to the home experimenter.

An excellent vegetable for observing browning is celeriac. It is possible to cut a large, relatively uniform, slice of this root and then lay several small filter paper discs on the cut surface, each soaked in different solutions such as lemon juice, vitamin C drink or apple juice. If the disc is carrying an agent that blocks the action of oxidases it will leave a white circle on the otherwise brown surface of the celeriac. Check the differences in browning prevention among the different solutions.

PS: The pH of a solution is a measure of how acid or alkali it is. Most substances fall on a scale of 1 to 14, with 7 being neutral. Distilled water is neither acidic nor alkaline so it registers a pH of 7, while acids, such as citric acid, fall between numbers 1 and 7 and alkalis, such as sodium bicarbonate, fall between 7 and 14. The more acidic a solution is, the lower its pH, while the more alkaline a solution is, the higher its pH.

Polyphenol oxidase was discovered in mushrooms in 1865 by Christian Schönbein, a Swiss chemist. It is found in humans, most animals and many plants. In plants its function is to protect against insects and microorganisms when the skin of the fruit is damaged. The dark brown surface formed by the skin is not attractive to insects or other animals, and the compounds formed during the browning process have an antibacterial effect.

In some foodstuffs made from plants this browning effect is desirable. For example, in tea, coffee or chocolate it produces their characteristic flavour. However, in other plants or fruits, such as avocado, apples and pears, browning is a problem for farmers, because brown fruit is not acceptable to consumers and, even more importantly, it doesn't taste good.

Bubble nuts

Can you wash your clothes with conkers?

Not all detergents come in packets; some grow on trees. Take horse chestnuts, for instance. When you're not threading them onto shoelaces, baking them in the oven or soaking them in vinegar to win conker championships, you could be using them to wash your whites whiter. But how can you clean your clothes with conkers?

What do I need?

- a handful of conkers
- a kitchen knife
- a chopping board
- a pan
- water
- a stove
- a tea towel
- a washing-up bowl
- a bottle with a screw top
- some dirty socks

What do I do? Remove the brown outer casing of the conkers, chop them up into small pieces, and put them in the pan. Add a cup or two of water and boil them for a few minutes, then let them cool. Strain the mix through the tea towel into the washing-up bowl to remove the solids and keep the liquid. Pour this into a bottle and shake it. Now put it back in the bowl and wash your socks in it.

What will I see? You should see a soapy lather form on the liquid you pour into the bowl. And, if you give your socks a good scrub, you'll see how clean they emerge.

What's going on? This highlights a property of conkers that few people are aware of. Horse chestnuts contain a saponin, a natural soap or surfactant. Many other plants, including soapwort, produce saponins, which help protect them against disease because they are toxic to bacteria and fungi. As we have seen, they can be extracted with water – a trick that has been used for centuries to make a soapy liquid for cleaning linen.

Surfactant molecules have a polar region that is attracted to water molecules (it is hydrophilic) and a non-polar region that is repelled by water (hydrophobic). They are therefore soluble in both water and organic solvents, including the substances that make your socks dirty. This makes them very similar to synthetic detergents and very good for cleaning.

Because saponins are very mild surfactants, they are less damaging to materials than stronger, synthetic detergents. This is why they are popular with art conservators, who use them to clean delicate fabrics or ancient manuscripts.

Surfactants also play a role in fire fighting. They are, along with a number of chemicals, a constituent of Light Water, a substance produced by the company 3M. As a dilute solution in water, this forms a light, stable foam used primarily to fight fires that involve flammable organic liquids such as oil. The foam floats on the surface of the burning liquid (hence 'Light Water'), covering it with a thin surfactant film that prevents further evaporation of the liquid and extinguishes the fire by cutting off the oxygen supply.

PS: Because this is a fun experiment which children enjoy, you need to be aware that the conker extract you have created is mildly poisonous if drunk, causing coughing and sneezing, and can be a skin irritant. So, close adult supervision of children is required, and when you've finished, give all the utensils you used a good wash to make sure there are no traces of saponin left.

Hot chocolate

Is it true you can measure the speed of light using nothing more than a chocolate bar and a microwave oven?

Yes, although a counter-theory says that all you need is a chocolate bar and a classroom of schoolchildren just waiting to pounce... This is a truly astounding experiment which actually allows you to measure one of the fundamentals of science – the speed of light – in your own home.

What do I need?

- a bar of chocolate (the longer the better)
- a metric rule
- a microwave oven

What do I do? Remove the turntable from your microwave oven – the bar of chocolate needs to be stationary. Put the chocolate in the oven and cook at high power until it starts to melt in two or three spots – this usually takes about 40 seconds. You should stop after 60 seconds maximum, for safety.

What will I see? Because the chocolate is not rotating, the microwaves are not evenly distributed throughout the bar and spots of chocolate will begin to melt in the high-intensity areas or 'hotspots'. Remove the bar from the oven and measure the distance between adjacent globs of melted chocolate.

What's going on? The frequency of the microwaves is the key here. A standard oven will probably have a frequency of 2.45 gigahertz (the figure should be given on the back of the oven or in the instruction manual). If your oven is 2.45 GHz, this means the microwaves oscillate 2,450,000,000 times a

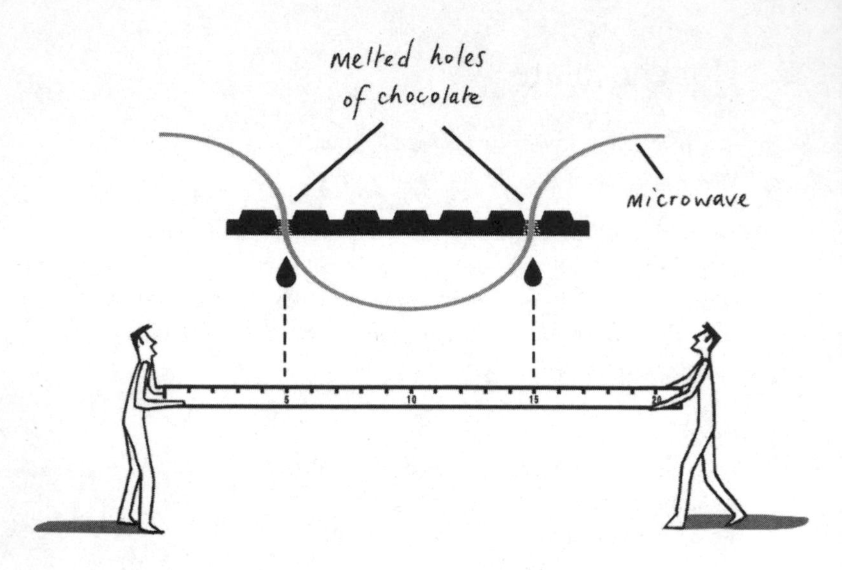

second (you can adjust this figure depending on your particular oven). Microwaves are a form of electromagnetic radiation and therefore travel at the speed of light. If you know the frequency of the microwaves, finding out their wavelength will help you to calculate how fast they are travelling.

This is where the chocolate comes in. The distance between the globs of molten chocolate is half the wavelength of the microwaves in your oven, so double the measurement you have taken of the gap between the molten globs to find the microwave wavelength. In the *New Scientist* microwave oven the distance between the globs of molten chocolate was 6 cm, so the wavelength in our 2.45 GHz oven is 12 cm.

To calculate the speed of light in centimetres per second you need to multiply this wavelength by the frequency of the microwaves: 12 × 2,450,000,000 = 29,400,000,000, which is astoundingly near to the true speed of light of 29,979,245,800 cm per second (or 299,792,458 m per second as it is usually expressed).

Try it yourself, measuring as accurately as possible to get a figure even nearer to the true speed. If your chocolate bar is chilled beforehand, the molten areas tend to be more distinct when they first appear. Of course, you may find a variety of different chocolate bars, all of which taste delicious slightly melted, will aid your research. True scientists know that it is always important to double-check results.

3 In the Study

(or the office or workroom ...)

Finger training

My guitar teacher says that I should train my fingers to be more dexterous by laying my palms flat on a table and wiggling each finger in turn. I do, but why do I find it more difficult to lift my ring finger than any of the others?

This is a more useful exercise than the old school classroom trick of getting your friends to place their knuckles on the desk in the form of a fist and extending each finger in turn and telling them you'll give them £10 if they can lift all their fingers off the surface one by one. They can't (the ring finger stays stubbornly on the desk), but both situations are affected by the same anatomical principles...

What do I need?

- a human hand with all fingers intact
- a flat, horizontal surface
- a £10 note that – all being well – you won't have to hand over

What do I do? Place your palm flat on the surface and try to lift and wiggle each finger in turn. Then place the second knuckles of your fingers on the surface, palm downwards in a fist, and try to lift each finger in turn, while keeping the knuckles of your other three fingers on the surface. Now, bet your friends £10 that they can't lift the ring finger from the surface...

What will I see? Your ring finger will be far less mobile than the others. And in the case of the knuckles-on-the-surface experiment, you'll find that, unlike your other fingers, you can't lift it at all.

What's going on? The anatomy of the human hand is very complicated. An 'extensor' muscle extends or straightens a digit when it contracts, while a 'flexor' bends it. Extension of a single digit requires the simultaneous contraction of the extensor muscle for the appropriate digit, the relaxation of the flexors for that digit, and the contraction of the flexors for all the other digits to make them stay where they are.

This happens fairly easily in the thumb, index and small fingers, because they all have their own designated extensor muscles. However, extension in the middle and ring fingers uses a common extensor muscle. When attempting to extend the ring finger, the middle finger flexor contracts as it is supposed to, effectively pinning down the middle finger, which the common extensor then pulls against to no avail. Try lifting the middle and fourth finger as a unit and see how easy this becomes when the middle finger no longer acts as a tether.

Because the hand evolved primarily for grasping, playing a musical instrument can push its design specifications to the limit. Researchers at Cardiff University studying the biology of musical performance have shown that there is a high degree of variability in the muscles and tendons of the hand and it seems that the success or otherwise of musicians can often be the result of differences in this physiology. Because of this, not everyone can make the same movements. For example, in about 20 per cent of people, curling up the thumb causes the index finger to bend inwards because of an anomalous link between flexor tendons. This renders some piano and guitar fingerings impossible for those affected.

On the other hand, evidence from touch-typists shows that their highly trained and flexible fingers can often move separately with no difficulty, while younger people who spend a lot of their time texting have astoundingly flexible thumbs…

The link between the middle and ring fingers is almost universal, but a very few people have aberrant connections, so if you can't afford to lose a tenner, beware of them and those other practitioners of finger flexibility, the touch-typists.

PS: *New Scientist* reader Conor Nugent has a different take on the immobility of the ring finger, telling us that 'the ring finger has the least dexterity because it is used the least. The index finger is used for pointing, the little finger is extended while drinking tea and the middle finger is used when driving. The ring finger, however, is used only once, shortly before marriage.'

See better

Why does looking through a pinhole allow you to see objects more clearly?

The author noticed this in his youth as he became progressively short-sighted and was trying to peer at the blackboard in classes and lectures. Sadly, he didn't find a similar cure for early-onset baldness. What he did discover was that if you peer through a small gap, such as that formed between curled-up fingers, objects that were blurred become clearer. The effect is particularly marked if you are short-sighted and remove your glasses or lenses before you try this out.

What do I need?

- the human eye (preferably a myopic one)

- a piece of card
- a pin
- paper with words written on it, much like an optician's eye-testing chart
- a comb

What do I do? Punch a tiny hole in the card with the pin. Fasten your home-made eye-testing card to a wall at a distance where you find it difficult to read. Peer through the pinhole in the card with one eye while keeping the other eye closed.

What will I see? The writing on the card will become clearer when you look at it through the pinhole. You'll notice the same when you look at other objects around you and in the distance. The effect is particularly noticeable if you are short-sighted, hence our earlier suggestion that if you normally wear glasses or contact lenses, remove them before attempting this. If you have excellent vision the effect will not be as obvious, but there has to be some compensation for short-sightedness, doesn't there?

What's going on? Under normal circumstances, light rays entering the eye's lens are not focused in one place. For you to see a clear image of a given point, the lens has to concentrate all the rays into a single point on the retina. When the eye is not perfectly shaped, as in short-sightedness or with an astigmatism, the outermost rays entering the eye are not bent or refracted by the lens enough for them to be focused on the correct point at the back of the eye. The innermost rays entering the eye do not need to be bent as much to hit the middle of the retina and these travel a relatively straight route to form a clear image, even in people who are short-sighted. However, the outermost rays confuse and blur this image. By looking through a pinhole the bundle of rays entering your eye is greatly reduced, because the hole allows through only

the inner rays that pass through the central portion of your lens straight to your retina, and excludes the peripheral rays which cause the blurring. This means you see images clearly again. The drawback to the improvement in vision is that the total amount of light entering your eye is greatly reduced by the pinhole, so images seem darker and, in low light, the reduction in brightness offsets the increased focusing gains.

Try using the gaps between the teeth of a comb to reproduce this effect. Native Alaskans have historically worn glasses with narrow slits in them, reproducing the effect of looking through a comb. More importantly, because snow and ice reflect a lot of light, which can produce excessive glare, looking through slitted glasses helps to reduce the amount of light entering the eye, aiding vision and preventing snow blindness.

PS: Pinhole glasses, similar to slitted glasses, use this effect to correct vision. They have a number of benefits over normal prescription lens glasses, the most important being that as your eye changes as you age, the pinholes do not need to be altered. One pair of glasses will last a lifetime. The obvious drawback, however, is that they greatly reduce peripheral vision.

Colour change

If you stare at a colour for long enough it seems to impinge on your vision. How?

Many of us have experienced this after looking at a bright light or after having our photograph taken when a camera flash was used. You spend the next five minutes or so blinking because you can still see the image of the light in your vision, even when you close your eyes. So why can you still see something that's no longer there?

What do I need?

- four large sheets of white paper
- pieces of bright red, blue and green card
- scissors
- glue
- a black marker pen
- a well-lit room

If you want to try the clever bit described at the end, you'll need:

- more A4 sheets
- green, yellow and black crayons

What do I do? Draw a fishbowl on one of the large sheets of white paper with the black pen. Cut out three identical fish shapes (large ones, but smaller than the bowl you have drawn), from each of the three coloured pieces of card and use the glue to stick these to the three remaining sheets of white paper, one on each. Hang all four sheets of paper on the wall of a brightly lit room.

What will I see? Stare at the red fish for about 30 seconds, then stare at the fishbowl. You should see a greeny-blue fish in the bowl. Look away, let your eyes return to normal, and repeat what you have just done with the green fish. This time you'll see a reddish-blue fish in the bowl. Repeat again with the blue fish and you'll see a yellow fish in the bowl.

What's going on? The retina of the human eye is covered with light-sensitive cells called rods and cones. The rods, of which there are more than 100 million, pick up on shading and levels of light and dark, while the cones register colour. There are fewer cones than rods, only around 7 million, but the cones are concentrated especially in the central area of the retina known as the fovea. There are three types of cone, each sensitive to a different range of colour – red, blue or green. Cones are more active at higher light levels, but rods, on the other hand, are responsible for vision in dusk and in the dark – hence the requirement for a brightly lit room to ensure this experiment works at its best.

The images you see in the bowl after staring at the fish are afterimages, but why do these appear in a different colour from the one you were looking at? When you stare at the red fish, the image falls onto one region of your retina and the cells responding to red become desensitised. White paper reflects all the colours of the spectrum rather than absorbing any of them, unlike the red, green and blue card from which the fish are made; instead these absorb colours, hence their appearance. So, when you turn to look at the fishbowl on the white paper, the red-sensitive cells don't respond as strongly as normal to the red part of the light that is being reflected back from the white surface, but the ones that are sensitive to blue and green light do. That's why you see a greeny-blue fish.

Obviously, if you stare at the green or blue fish before looking at the white paper, the green- or blue-sensitive cells become fatigued and the fish in the bowl changes colour accordingly.

But what causes the afterimage in the first place? The old explanation was that the cone cells become tired when they have to constantly 'fire'. In fact, the afterimage is due to the nerve cells in the eye that process the signals from cone cells. When they keep receiving a red signal from the cones, they start to adapt by turning down the 'volume' of the signal they send to the brain.

When we look away from the image, the red signal from this part of the eye stays muted for a while. However, the blue and green signals are sent at the normal 'volume'. Your brain doesn't appreciate that the red signal has been turned down and interprets this as a normal signal from your eyes. This means that the fish-shaped patch from which the red signal is reduced is perceived as a genuine object in your field of vision, leading to the afterimage which persists even when the object has been removed.

Normally the eye deals with the problem described above by shifting slightly, so that the same cones are not constantly exposed to the same image, and there is no desensitisation. However, if the image is large enough, like the fish you have been staring at, the small eye movements are not enough to change the colour that is being detected by one part of the eye, and, as we have seen, this part will stop responding as strongly as normal.

PS: A clever way to test this phenomenon is to draw a Union Flag on a piece of A4 paper, replacing all the red with green, all the blue with yellow and all the white with black. Stare at the flag for 30 seconds and then stare at a blank white background as before. The flag will appear as an afterimage in the correct colours.

Sealed in light

Why do self-sealing envelopes appear to spark when opened in a darkened room?

It's important to make sure you use a self-sealing, pre-gummed envelope to see this. The good news is that it will work repeatedly on the same envelope – just reseal it after opening and pull back the flap again.

What do I need?

- self-sealing envelopes
- a darkened room

Sugar cubes and a pair of pliers can be used to create a related effect.

What do I do? Draw the curtains and switch off the lights (preferably at night). Let your eyes adjust to the darkness. Pull open some self-sealed envelopes, making sure the two gummed and sticky pieces of paper keeping the envelope closed are pulled apart as rapidly as possible without tearing the rest of the envelope.

What will I see? A fluorescent glow, or 'sparking', will appear along the line of the gummed surfaces as they are torn apart.

What's going on? The coloured glow is a form of chemiluminescence – the emission of light without heat as a result of a chemical reaction. Separating the gummed surfaces requires energy that breaks the attractive forces between the molecules of gum.

The act of pulling apart the surfaces supplies excess energy to the gum molecules, which lifts them into an excited state. As they decay back to their normal state, energy is released in the form of visible light. The difference in energy between the excited and ground states defines the wavelength and hence the colour of the light produced; in this case purple.

Chemiluminescent glow can also be seen when stripping lengths of electrical insulating tape, and although people have speculated that it might be responsible for some gas-induced mining explosions, there is no evidence to suggest there is enough energy in the sparks to ignite the methane present in mines.

This phenomenon is different from fluorescence, where light (often ultraviolet light) is absorbed and then re-emitted at a longer wavelength (in the visible spectrum). Fluorescence gives rise to loud, brash, glowing colours such as the kind of blue glow you might observe while drinking tonic water near one of the ultraviolet lamps often found in nightclubs.

PS: You can see a similar effect if you take a pair of pliers and crush sugar cubes in a darkened room. When sugar crystals are crushed, tiny electrical fields are created, separating positive and negative charges that then create sparks while trying to reunite. In this case the phenomenon is known as triboluminescence – light generated by the breaking of bonds in a crystal.

Elastic fantastic

Why do rubber bands heat up when you stretch them?

The next time someone flicks a rubber band at you, don't shoot it back at them – at least, not straight away. First, take a moment to try a simple experiment which reveals a strange behaviour of this everyday object.

What do I need?
- a thick rubber band
- a hairless upper lip
- a skipping rope
- a friend (if you want to study the effect further)

What do I do? Stretch the rubber band between your hands and extend it as far as it will go. Now hold it against your upper lip and let it contract quickly. Wait a moment, then, keeping it against your lip, quickly extend it again.

What will I feel? You'll notice that as you let the rubber band contract it feels cool. As you extend it the rubber feels hot. Quickly release the band and it will feel cool once more.

What's going on? Rubber is made up of a tangled network of extremely long, flexible polymer chains – a bit like a crumpled fish net – that are constantly vibrating and rotating.

Polymers are substances with molecular structures built up mainly from a large number of similar units bonded together.

As you stretch a rubber band, these polymer chains align and are pulled closer together, reducing the range of motions they can undergo. This reduces their entropy (the degree of disorder or randomness they possess). So, according to the laws of thermodynamics, to maintain equilibrium, entropy has to increase in some other way to counteract this, and the atoms in the chains compensate by vibrating more vigorously. As a result, the rubber heats up.

When the rubber is allowed to contract, the process reverses – the chains become free to rotate and vibrate, and their entropy or randomness increases. To compensate, the entropy of the atoms in the chain decreases, the atoms vibrate less and their temperature falls.

Now you can shoot the rubber band back…

PS: Rubber's unusual structure means it behaves quite differently from other materials. If a length of stretched rubber is heated, for example, its polymer chains will begin to vibrate more vigorously and this tends to pull the ends of the chains towards each other – in other words the rubber will contract. This is the opposite of the usual thermoelastic effect exhibited by metals, for example, which usually expand when heated.

The effect is easy to demonstrate. Hold one end of a skipping rope and ask a friend to hold the other end. Begin to turn it as you would if somebody was skipping. You'll notice that the faster the rope turns, the harder it is to hold the two ends apart. This is how rubber behaves, pulling in the ends of its polymer chains as it vibrates more vigorously under heating.

Rubber horror

Why, when left for long enough, do rubber bands seem to melt and turn into a sticky mess?

After trying out the previous 'Elastic fantastic' experiment, put the rubber bands aside for a few months, preferably in a sunny place, and see how they change.

What do I need?

- rubber bands
- a sunny indoor spot such as a window ledge or table near a window
- greaseproof paper to protect your chosen surface

What do I do? Leave the rubber bands on the greaseproof paper for a few weeks or months in full sun.

What will I see? Over the weeks and months the bands seem to melt, at first becoming a sticky mess and then, when left even longer, often becoming brittle and crumbly.

What's going on? Natural rubber is made up of polyisoprene chains that slip past each other when the material is stretched. When raw, the substance is too sticky and soft to be of much use, so it is toughened with chemicals such as sulphur. These create cross-links between the chains, making the rubber stiffer and less sticky. This process is called vulcanisation.

Over time, ultraviolet light and oxygen in the air react with the rubber, creating reactive radicals that snip the polyisoprene chains into shorter segments. This returns the rubber to something like its original soft and sticky state – what you see when your bands have turned gooey. Later, the radicals form new, short cross-links between the chains. This hardens the rubber and eventually turns it brittle, leading to

the crumbly end product. Any vulcanisation agents left in the rubber contribute to this process.

Whether a rubber band is going sticky or brittle depends on the relative rates of these processes, and these rates in turn depend on the rubber's quality – what additives, fillers and dyes it contains and whether these absorb light or help transfer radicals – and how it is stored. Heat and light speed up the reactions, and the presence of strong oxidisers in the air such as ozone creates even more radicals. The eventual fate of your rubber band depends on the room temperature and whether you have a desk by a window or next to an ozone-emitting machine, such as a photocopier.

How much heat and light is required for these changes? The polymer chemistry of rubber is fairly complex and so this is difficult to answer precisely. Obviously, the chemical reactions take longer if the rubber bands are left in a fridge than on a sunny window ledge. A rule of thumb is that reaction rates roughly double for a 10 °C rise in temperature, but this is more complicated when you take oxygen and light into consideration.

The final factors that influence the change are ozone concentration, ultraviolet light intensity and whether the band is stretched or not – stretching brings the polyisoprene chains closer together, allowing radicals to jump from one chain to another more easily and so create new bonds between the chains.

PS: The biggest consumer of vulcanised products is the automobile industry, which purchases almost 80 per cent of all new rubber, mainly to produce vehicle tyres. The world produces more than a billion tyres annually, with an estimated 3 billion in use in Europe and 6 billion in North America at any one time.

Rip off

Why does newspaper have a preferential direction for tearing?

It's true, newspaper looks homogeneous, but when you try to tear it from top to bottom or from side to side, there is a noticeable difference. And newspaper isn't the only everyday object that has a grain…

What do I need?

- a newspaper, preferably a broadsheet to give you more ripping length

If you are happy with the results of your experiment you can also buy some uncooked hamburgers (a mixture of mass-produced and posh ones helps to display the difference in types) and hang on to the plastic bag you carried them home in.

What do I do? Rip one sheet along its length, then rip another sheet along its width.

What will I see? One of the pieces will, with little difficulty, tear parallel to the edge of the paper. The other will rip haphazardly.

What's going on? Most paper is made on machines which operate at high speed. The sheets are made by draining a dilute suspension of fibres and mineral fillers on a table of continuous and fast-moving, synthetic, sieve-like wire. The sheets are then consolidated by pressing and finally dried on heated cylinders before being put onto reels.

As the paper-making suspension is discharged from a vessel known as the flow box or head box onto the rapidly moving wire, most fibres – which are cylindrical in shape – are aligned in the direction of the wire's movement. This is

called the 'machine direction'. The orientation of fibre in the paper structure allows the sheet to tear more easily in the machine direction than in the 'cross-direction' where the fibres are presented sideways.

The strength of all paper made in this way is influenced to varying degrees by the directional properties of the sheet. The addition of fillers and mechanical or surface treatment during manufacture tends to modify or reduce this. Newsprint tears more easily in the machine direction than computer or photocopier paper, as it contains mostly fibre and is lighter in weight.

PS: You can look for a 'machine direction' in any number of household items and foodstuffs – many possess similar qualities. Hamburgers cooking at the *New Scientist* summer barbecue were seen to change shape from round to oval due to similar manufacturing circumstances.

Commercial hamburger patties are made on a large scale – their ingredients are minced, extruded, rolled out, flattened and stretched on a moving belt to give them the right thickness and texture before the patty discs are punched out by machine. It is this extrusion and rolling that makes the hamburgers anisotropic (meaning their physical properties have different values when measured in different directions), because their protein fibres tend to get stretched along the direction of the belt of minced meat.

When cooking starts, the heat makes the protein expel water and shrink. The stretched fibres mainly shrink lengthways and this shrinkage squeezes the circular patty more lengthways than across the direction of the moving belt, and therefore shortens them into an elliptical shape – until you eat them. By contrast, handmade hamburger patties are isotropic (meaning their physical properties have the same values when measured in different directions) and cooking them will not alter their shape in any particular direction.

Many other rolled or extruded products display anisotropy. The plastic used to make carrier bags is extruded and stretched and so tends to have a preferred tearing direction. The bags are designed so that when loaded vertically and carried by their handles they are at their strongest.

Immobile mobiles

Does talking on a mobile phone really slow your reactions?

It's a hot topic right now because in the UK and elsewhere driving while talking on a hand-held mobile phone has been made illegal. Sceptics seem to think that such legislation is an overreaction and often you'll see people flouting the law. Here's a way to judge for yourself just who is right.

What do I need?

- one or more experimental subjects (preferably people who think their driving is unaffected by talking on a mobile phone)
- a yardstick (any rod-like instrument that is divided into 1 cm increments will do – you can probably even use a rule as long as it is at least 90 cm long)
- a table
- a calculator

What do I do? Ask the subject to place their forearm on the table with their open hand overhanging the edge, and with index finger and thumb at the top as if they were about to shake hands. Hold the yardstick vertically just above their hand, with the bottom of the yardstick level with the index finger. Ask your subject to catch the yardstick as quickly as possible after you drop it. Note down how many centimetres have slipped before it is caught. Give them three tries and take the average. Now ask your subject to call a friend on the phone and enjoy a good natter. When the conversation is in full swing, drop the yardstick. Now see where it is caught.

What will I see? You will see a crude measure of how a telephone conversation can affect reaction time. And you'll find that chatting to a friend can seriously harm your ability to respond to a change in circumstances.

What's going on? Most researchers who study multitasking agree that holding a conversation interferes with a person's ability to react to events, especially those occurring on the road. When two stimuli are presented simultaneously the reaction to one of the tasks will be delayed. And of great significance to all road users – drivers and pedestrians alike – is that you can't always choose which. So when your friend asks you something that begs a reply and a child steps into the road in front of you, there is a moment's hesitation as your brain attempts to make sense of the situation and select which is the more important issue to act on. This is known as the brain's 'response selection bottleneck' and was first noted by researchers in 1952. The decision process may seem short, but at the speed a car travels it can easily make the difference between life and death.

PS: To get the person's reaction time in seconds, divide the number of centimetres recorded by the falling measure by 490 and take the square root. Results taken from a number of volunteers in the *New Scientist* office provided us with average figures of 17 cm when volunteers were concentrating and 41 cm when they were being distracted by somebody talking to them on the telephone. These distances correspond to reaction times of 0.186 seconds and 0.289 seconds respectively, meaning people on the phone took 0.103 seconds longer to react – an increase of 55 per cent.

This gives an idea of how people's reactions are delayed when they are talking on the phone in a quiet office, but on the road there is even more to distract a driver, including having a conversation with a fellow passenger or listening to the radio, both of which can increase the delay. Because a car travelling at 100 km an hour (about 60 miles per hour) travels almost 28 m (92 feet) in 1 second, the delay noted in our example could be crucial. The distance the car travels in this time is simply its speed multiplied by the time. In this case this would be 2.86 m – nearly a car length, and possibly the difference between life and death.

Air space

Why, when a candle is standing in water and is covered by an upturned glass, does the water in the glass rise when the candle goes out?

This question was originally asked in *New Scientist* by three Australian schoolchildren, Emma, Rebecca and Andrew Fist of Norwood, Tasmania, who wanted to know why if they used four candles under the glass instead of one, the water rose even further up the upturned glass. They decided that what they had been told by their science teacher couldn't be correct, and wanted to know why…

What do I need?

- four candles
- a large glass or jar capable of covering all four candles after they have been lit
- a bowl of water with a flat base

What do I do? Stand a single candle in a bowl firmly anchored to the bottom so that it can't topple over. Fill the bowl with water to a few centimetres depth. Light the candle. Place the upturned glass over the candle. Repeat the experiment using four candles under the upturned glass. You can do this

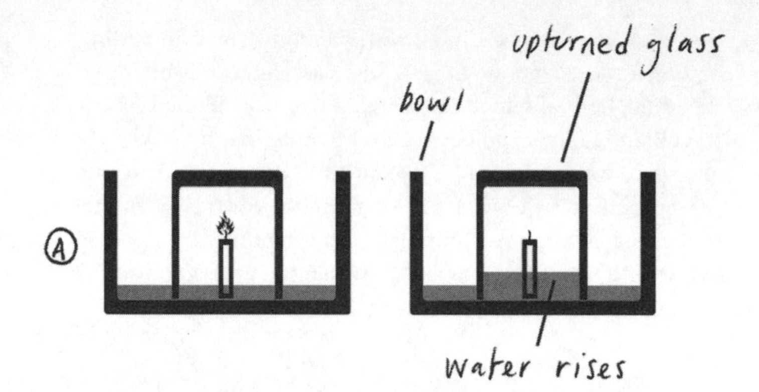
upturned glass
bowl
A
water rises

B

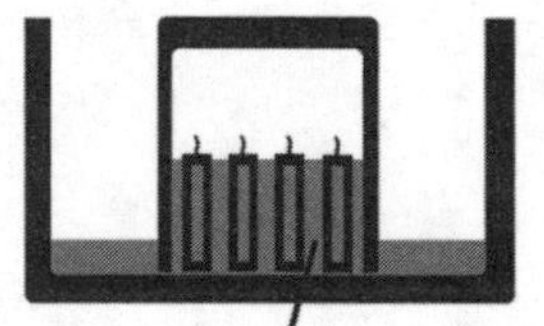
water rises higher with four candles

C

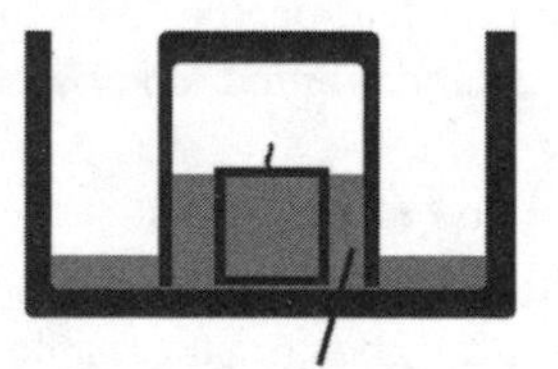
water can rise as high as in diagram B if a single thick candle is used

experiment in conjunction with the 'Shedding light' experiment on page 24 because they use similar equipment.

What will I see? When the candle goes out, the water level in the glass rises. Intriguingly, when you repeat the experiment using four candles, the water rises even higher than it did in the single candle experiment.

What's going on? It will come as a shock to many schoolchildren who were shown this experiment as part of a school science lesson – and it might surprise a few teachers too. Many of us were taught that the rising water level is caused by the oxygen under the glass being consumed by the burning candle and the water rising into the glass to replace the void created by the missing oxygen.

However, it took three schoolchildren to question the orthodoxy associated with this classic experiment.

The consumption of oxygen by the burning flame may well contribute to the rising water level to a certain extent because a given volume of oxygen will burn the candle wax's carbon into roughly the same volume of carbon dioxide and the wax's hydrogen into twice the volume of water vapour respectively. The former will partly dissolve into water, the latter will almost completely condense into liquid water. This will lead to a net decrease in gaseous volume.

However, this is a minor consideration. The important influence in what you see is the heat created by the burning candle or candles. When they are covered with an upturned glass, several candles will increase the air temperature around themselves more than a single candle will.

As soon as the candle or candles go out, the surrounding air contracts as it cools, allowing more water to enter the glass, and the ratio of contraction is directly proportional to the initial average temperature of the air volume under the glass. More candles lead to more heat, a higher temperature before they are extinguished, and therefore a higher water level upon cooling. The fact that there is a disparity in water depth between a single candle and four candles shows us that the absence of oxygen cannot be the key factor, because the same amount of oxygen is consumed in each case. The reason that water rises into the jar is due to a contraction of the cooling gases after the candles have been extinguished.

Further proof that contraction of expanded gases is the reason for the rising water level comes from the fact that the water level only rises after the candles have been extinguished, at which point the remaining gases cool. If the water rose because the oxygen under the glass was being consumed it would rise in a linear fashion throughout the period of burning.

Incidentally, a candle flame goes out after only a small percentage of the oxygen has been used up, because oxygen must be at its usual proportion (it forms 21 per cent of air) to allow flames to burn freely. Any less and burning is quickly retarded. The rise in water level is far too great if the water was replacing only the missing oxygen.

One other factor must be taken into account. The effect is partly caused by the thickness of the three extra candles in the second version of the experiment, and you can get a similar effect as this four candle version using a single candle of greater thickness. The thicker the candle, the higher the water will rise. The water drawn into the glass is squashed between the candles and the glass and, obviously, the narrower and smaller this space, the higher the water will rise.

Sucking eggs?

We know why water is drawn into a jar after the candles go out, but how does a similar process end up with an egg in the bottle?

This is a classic experiment, repeated many times down the years. We include it here because so many people mentioned it after *New Scientist*'s readers debunked the candles-under-the-jar experiment.

What do I need?

- a hard-boiled egg
- a bottle with an opening slightly smaller than the circumference of the egg when it is placed upright atop the bottle (milk bottles are ideal; wine bottle necks tend to be too narrow)
- long matches

Vinegar and syrup will allow you to carry out a second experiment.

What do I do? Shell the egg so you have the fresh, rubbery white exposed. Place the bottle on a flat, flame-resistant surface. Light a match and drop it into the bottle, making sure it stays alight. Add a few more burning matches as quickly and as safely as possible – if you can, add all the matches at once. Put the egg, pointed side down, over the neck of the bottle.

What will I see? The matches will soon go out, because they are starved of oxygen once the egg seals the bottle, and the egg will be pushed into the bottle, looking rather peculiar as it contracts in order to slide down the inside of the neck.

What's going on? The lighted matches heat up the air inside the bottle causing it to expand. This means that some of the expanded air escapes from the bottle. When the matches go out after the egg has been placed on top of the neck, the air contracts, creating a lower pressure inside the bottle than outside. The greater pressure outside the bottle forces the egg inside and it plops to the bottom.

While it seems as though the egg is being pulled or sucked into the bottle, in reality it is actually pushed by the greater pressure outside the bottle. The air outside the bottle, at greater pressure than that inside, forces the egg downwards. An egg works well because its moist, rubbery surface forms an effective seal around the neck of the bottle. If you repeat this experiment with a chocolate egg or an unpeeled, uncooked new potato that doesn't completely seal the neck of the bottle, air will be able to return to the bottle without carrying the object with it.

You can get the egg out again by holding the bottle on its side and carefully jiggling the egg so that its narrow end is resting against the bottle's neck. Turn the bottle upside down and form a seal between your mouth and the bottle opening. Now blow hard, but be careful, the egg can come out at quite a speed.

In similar fashion to the mistakes made when solving the candles conundrum (see the previous experiment, 'Air space'), many people were taught that the egg is sucked into the bottle when the burning matches use up the oxygen in the air inside the bottle. Wrong again…

PS: While you have your eggs out, you can check out another experiment, especially if you've saved some vinegar after trying out the 'Plastic milk' experiment on page 70.

Place a raw, unshelled egg in a jar of vinegar and leave it for a couple of days. When you return you'll see the shell has disappeared and the egg has swollen to perhaps twice its size. This is because the shell, which is made of calcium carbonate, reacts with the vinegar's acetic acid creating carbon dioxide, water and calcium ions. The water passes into the egg through its now exposed membrane, driven by the pressure of osmosis (the passage of water from a region of high water concentration through a semi-permeable membrane to a region of low water concentration, reducing the difference in concentrations). The white of the egg has a very high concentration of protein, and water passes from the vinegar into the egg in an attempt to equalise the concentrations on either side of the membrane.

To reverse this process, place it in a solution of 75 per cent sugar syrup and 25 per cent water. After a couple of days the egg will shrink to less than its original size as osmosis works in the opposite direction and water leaves the egg to equalise the concentration between the egg and the sugary syrup mix surrounding it.

Cloud burst

Clouds can be created in a plastic bottle. How?

What's even more interesting, you can make them vanish and reappear just by squeezing the bottle.

What do I need?

- matches
- water
- a clear, flexible 2-litre plastic bottle with a screw cap

What do I do? Add just enough water to the bottle to cover the base, and shake it around. Light a match, let it burn for a couple of seconds then blow it out. Immediately drop the smoking match into the bottle and screw the lid on quickly and tightly. Now squeeze the bottle hard four or five times.

What will I see? A cloud will appear in the bottle after you have added the match. But when you squeeze the bottle the cloud disappears. Release the bottle and there is the cloud again ...

What's going on? In effect you are building a small cloud chamber. The cloud in the bottle forms as a result of three factors: the smoke from the match, tiny drops of water inside

the bottle formed after you shook the bottle when the water was added to it, and the change in air pressure caused by squeezing or releasing the bottle.

Water in the bottle combines with air to form water vapour – the gaseous form of water. This gives rise to the cloud in the bottle. But for water vapour to form this cloud, particulates also need to be present. In this case smoke provides the necessary particles, which act as nucleation sites, allowing drops to collect on them. Without the smoke no cloud will appear.

Squeezing the bottle raises the pressure. This increases the temperature inside the bottle, helping the water to change from visible liquid back to invisible gas (most liquids turn into gases as temperature increases), and the cloud disappears. Releasing the pressure reverses the effect.

PS: Real clouds form in exactly the same way. In this case the water vapour comes from the evaporation of seawater, rivers and lakes. This expands and cools as it rises. There is a limit to the amounts of water vapour the air can hold and this is higher when the air is warmer. As the temperature drops at higher altitudes the vapour begins to condense and clouds are formed, just as in the bottle. Clouds form when the bottle is left alone, which represents a cooler atmosphere, but they disappear when the bottle is squeezed and the pressure and temperature are raised. This is exactly what happens in the atmosphere. At higher altitudes the gas expands, is less compressed and is cooler, so clouds appear. At lower altitudes the gas is warmer and more compressed, so there are no clouds. And, because the atmosphere contains many small particles, ranging from dust to smoke or salt particles, there are plenty of nuclei around which condensation can gather to form clouds if the pressure and temperature conditions are correct.

Spark of genius

How does an electrophorus work?

An electro-what? This may sound like something the Daleks would have used for world domination in the early days of *Dr Who*, but it's actually a simple device that demonstrates, quite spectacularly, the effect of stored electrical energy, and you can make your own version at home.

What do I need?

- a polystyrene plate
- wool (you can use a jumper, but a carpet has been known to work)
- a flat-bottomed, aluminium foil dish (the kind used to hold pre-packed quiches, pies or puddings)
- a pencil with a rubber at the end
- a drawing pin
- superglue

What do I do? Press the drawing pin through the bottom of the dish so its point is sticking up into the middle of the inside of the dish. Press the rubber-ended pencil down onto the point of the pin so that the pencil juts – point upwards – from the base of the dish. A little superglue will help to hold it firmly in place. You should be able to lift the whole assembly using just the pencil. If you can't get this to work, glue a polystyrene cup to the foil dish and use that to lift it.

Now rub the polystyrene plate with the wool for at least three minutes. Place the plate on a table or work surface and, holding only the pencil handle, position the aluminium dish on top of the polystyrene plate. Touch the dish once with your finger while it sits on the polystyrene plate. Now lift the dish off the plate, again using only the pencil, and touch the dish again.

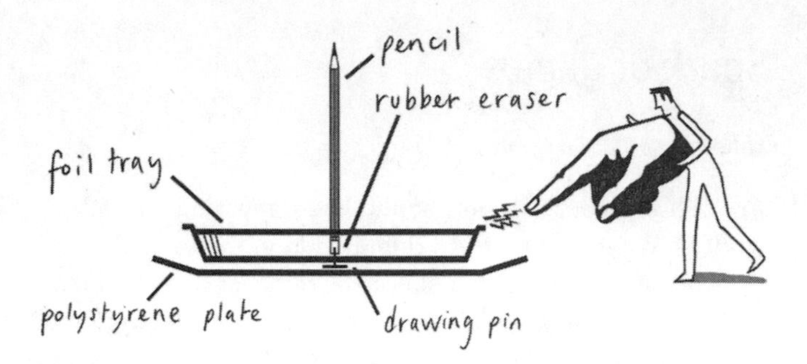

What will I see? The first time you touch the dish (i.e. before removing it from the polystyrene plate) a spark may fly from the dish to your finger, and you'll certainly experience a small electrical shock. When you remove the dish from the plate and touch it again with your finger, you'll see another spark and really feel the electricity jump from the dish to your skin. If you want to see the sparks produce spectacular results, try the experiment in a darkened room.

What's going on? Rubbing the polystyrene plate with wool creates a negative static charge on the surface of the plate by attracting electrons (which are negatively charged). This is called triboelectrification. By placing the aluminium foil dish on the plate using the rubber-ended pencil you are ensuring that no charge is lost through the dish because it is not earthed through you – you are holding the wooden part of the pencil and the pencil is stuck to the dish by an insulating rubber. When the aluminium foil dish makes contact with the polystyrene plate its electrons are repelled from its base because the negative charge on the electrons in the polystyrene, created by the wool rubbing, repel the negatively charged electrons in the foil dish. This is because like charges repel each other in a similar way to two magnets.

This makes the bottom of the aluminium foil dish positive and the top of the dish negative. So, by touching the foil dish with your finger while it is still resting on the polystyrene plate, you allow these negative electrons that have been forced to the top of the dish to flow into you, creating a spark and an electrical shock that you feel in your finger. Because you have removed the negatively charged electrons from the foil dish, the dish is now positively charged. If you remove it from the polystyrene plate using your insulated pencil handle and touch its positively charged surfaces again with a finger of your other hand while it is suspended, you'll create another spark and a shock as the foil dish is made neutral again by your earthed body.

PS: You have made a device called an electrophorus, perfected in 1775 by Alessandro Volta, who was responsible for the development of the electric battery and after whom the electrical volt is named. The electrophorus was adopted by physicists around the world because it provided a source of charge and voltage for research into electrostatics.

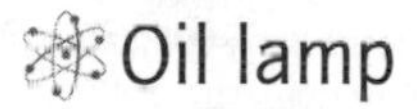

Oil lamp

Alka-Seltzer tablets can be used to create a 1970s-style lava lamp. How does it work?

This is a trendy effect – or it is if you grew up in the 1970s listening to recordings of Woodstock and wearing love beads. It works with any effervescent antacid tablet…

What do I need?

- a clean plastic 2-litre bottle
- water
- food colouring
- vegetable oil
- Alka-Seltzer or other effervescent antacid tablets

What do I do? Colour the water with the food colouring – about 10 drops ensure the water will be quite dark. You can choose the colour of the water, but red or purple gives a pleasing effect. Fill the bottle three-quarters full with vegetable oil, and top it up with the coloured water. Divide the antacid tablet into eight and drop one piece into the bottle.

What will I see? Oil and water don't mix, so at first the oil – which is less dense than the water – will float above the coloured band of water at the bottom of the bottle. You'll also notice that the food colouring stays in the water and doesn't taint the oil. When you add the tablet it will fall through the oil to the water below and begin fizzing as it reacts with the water. Then you'll see globules of coloured water begin to rise up through the oil before reaching the surface and sinking again. When the reaction stops you can start it again by adding another piece of antacid tablet.

What's going on? Antacids are made from sodium bicarbonate and citric acid. When these are placed in water they react vigorously. One of the key products of this reaction is carbon dioxide, plus water and various salts. So when the tablet reaches the water at the bottom of the bottle it reacts with the water, forming bubbles of carbon dioxide. These bubbles become attached to the oxygen and hydrogen molecules that make up water and, because the bubbles are much lighter than water or oil, they rise to the surface, dragging coloured liquid with them. Oil is viscous, so their journey to the surface is relatively slow, which means they rise with a pleasingly gentle action. When they reach the surface they burst, releasing the coloured water they dragged along for the ride and allowing it to sink once more to the bottom of the bottle. It is this constant rising and falling of coloured water that leads to the impressive lava lamp effect.

PS: The original lava lamp was invented by Ronnie Rossi in the 1960s. It comprises a glass bottle sitting atop a light bulb that heats the contents of the bottle. Inside the bottle are water and coloured wax with a metal coil at the base. At room temperature wax is denser than water, but when it is heated it becomes less dense. So when the lamp is switched on and the bulb heats the bottom of the bottle and its contents, the wax, which at room temperature sits at the bottom of the bottle, begins to heat up. As it becomes less dense, globs of wax begin to rise through the water. When a globule rises, however, it cools as it moves away from the heat source, and its density begins to increase again. This means it falls back through the water towards the bottom of the bottle, where it starts to heat up again. The metal coil also helps the wax to recoalesce so that the process can restart. The separated pieces heat up again and rise once more, creating a cycle as long as the heat-supplying light is switched on. A temperature of around 60 °C needs to be maintained at the bottom of the bottle containing the wax and water. If the system is too cold, the wax will simply sit at the bottom of the lamp; if it's too warm, it will float on the surface.

Seeing double

If you place one mirror facing another and stand between them, can you see infinity?

This question is perhaps more philosophical than scientific, but there's no doubt that when you look into one or other of the mirrors your reflection seems to stretch away into the far distance. So does it actually have an end?

What do I need?

- two mirrors, the bigger the better

What do I do? Place the mirrors facing each other as parallel as possible – standing them along two facing walls of a room will help you align them better. Stand between them and look at your reflection in either mirror.

What will I see? Your reflection will seem to stretch into the distance in both mirrors, curving away to become smaller and smaller until you can no longer determine where it ends.

What's going on? Sadly, you can't see infinity. No mirror reflects 100 per cent of the light falling on it. If it's a very good mirror and can reflect 99 per cent of the light, after about 70 reflections only 50 per cent of the light will remain, after 140 reflections only 25 per cent of the light is left, and so on until there is not enough light left to reflect between the two mirrors. Additionally, most mirrors reflect some colours of light much better than others and some colours are absorbed better by the glass, so the multiple reflections that you see not only get darker but they also become more colour-distorted as they recede.

Then there is the problem of mirror alignment. Even with perfect reflection of all colours you could never see infinite reflections, for geometric reasons. The faces of the two mirrors would need to be perfectly parallel. This is practically impossible to achieve. There will always be a slight disparity in their positions relative to one another. This is why the images appear to curve away, until eventually the reflection is lost 'around the bend'.

Even if the mirrors are perfectly reflective, perfectly parallel and really huge, your eyes are in the middle of your head, not at its edge. Therefore, at some point, the receding and hence apparently more distant mirror images would become smaller than, and hidden behind, the first reflected image of your head. Even with a tiny camera and a giant pair of mirrors, the reducing reflection size would eventually be smaller than the first reflected image of the camera apparatus.

However, for those who prefer a theoretical mathematical explanation of whether infinity awaits us at the end of a reflection, there is hope. Assuming perfectly parallel mirrors, perfect reflection and a transparent viewer who was prepared to stand around for ever, infinity is just about achievable.

Take a deep breath – here's how it works. Light travels at a finite speed (*c*), which you measured in the 'Hot chocolate' experiment on page 79. (In case you haven't done that experiment, light travels at roughly 3×10^8 m per second.) If your two mirrors were *L* metres apart and you stood between them for *t* seconds you would be able to see *c* multiplied by *t* divided by *L* reflections. For example, if your mirrors were 2 m apart and you stood between them for 1 minute, you should be able to see about 9 billion reflections. If you could stand around for ever, those reflections would presumably reach infinity…

PS: If you are short-sighted you can take advantage of having your mirrors set up to study another interesting property of mirrors. You may have to shift them about a bit to get them in a position where they are reflecting a distant object. Then take off your glasses or remove your contact lenses and you'll notice that although you are standing close to the mirror, the distant object that is being reflected is as blurred as if you were looking straight at it. This may seem odd because you are very close to the mirror.

The reason the object is still as blurred as if you were looking directly at it is because the light still has to travel the same distance (plus a little bit more) to your eye. It travels from the object to the mirror and then onto your eye, so you are looking at the sum of the two distances, not at the mirror itself.

A simple experiment to prove this is to get a small sticker and place it at eye level on the mirror. You'll notice that looking at your reflection, looking at the reflection of an object behind you, and looking at the sticker on the mirror all require you to adjust the focus of your eyes accordingly. It appears the object behind you is actually behind the mirror, and exactly that same distance away from it. This proves that your eyes do not focus on the surface of the mirror but on the objects reflected in the mirror and the distances these objects are from the mirror.

Anybody who is acquainted with photography will understand these principles. When using a camera to focus on an object in a mirror, photographers have to calculate the effective distance by measuring the distance from the camera to the mirror and adding it to the distance from the mirror to the object. Some autofocus cameras have particular trouble with this process because they only read the distance from the camera to the mirror and so need to be manually adjusted to compensate for this.

There are lessons to be learnt from this if you are in the habit of checking your appearance in the mirror. Remember that you are looking at yourself from twice the distance to the mirror. If you are checking for blemishes or stray pieces of hair you need to halve the distance between yourself and the mirror to see yourself as others do.

4 In the Bathroom

The stuff of life

I've been told you can extract your own DNA at home. Is this true?

Of course it is. And, as seems to be the case in so many *New Scientist* experiments, this involves a stiff drink. However, there's a perfectly good reason for breaking out the whisky – you can actually find out what makes you unique in the comfort of your own home...

What do I need?

- a teaspoon of salt dissolved in a glass of water
- a small, clean glass
- some washing-up liquid
- an eyedropper
- an ice-cold spirit drink of greater than 50 per cent alcohol by volume (cask-strength malt whisky, a quality gin, strong vodka or rubbing alcohol should do the trick)

What do I do? Put a teaspoon of washing-up liquid diluted with three teaspoons of water into the clean glass. Swish the salty water around your mouth vigorously for 30 seconds or so then spit it into the diluted washing-up liquid. Stir this firmly for a few minutes, then very gently pour a couple of teaspoons of ice-cold strong alcohol down the side of the glass. Use the eyedropper if you don't have a steady hand; tilting the glass also helps. This stage requires great concentration and is very important as you must have a clearly demarcated water/alcohol boundary. If you are careful, the spirit will form a separate layer on top of the salt/spit mix.

What will I see? Wait a few minutes and you'll see spindly, white, thread-like clumps starting to form in the alcohol. This is your DNA.

What's going on? Swishing salty water around your mouth removes cells from the inner surface of your cheeks in the way you've seen in TV dramas when the police take a swab from a suspect's mouth for DNA analysis. The detergent in the salt/spit mix breaks down the cells' membranes, releasing the DNA in the cell nuclei. Because DNA is soluble in water but not in alcohol it precipitates out in the white clumps you see floating on the surface. If you have a microscope you can investigate further, but if you just want to sit back and admire it, that stuff floating in the glass is what makes you who you are…

PS: Because the spirit has to be chilled, malt whisky aficionados may turn their noses up at putting their favourite spirit in the fridge, but vodka lovers will be happy enough to contribute to this experiment. The whisky drinkers will either have to swallow their pride, buy a bottle of vodka or gin, or simply slum it with rubbing alcohol.

Before you start make sure you have a clean mouth – if you've been scoffing a meat sandwich, it might not be your own DNA that ends up in the glass.

Testing the waters

Can we trick our sense of touch?

This popular experiment shows us that much of what we perceive through our senses is relative. We adapt to our changing circumstances.

What do I need?

- three bowls large enough to put your hands in
- a decent depth of hand-hot water, tepid water, and cold water chilled in the fridge
- your hands

What do I do? Simultaneously place one hand in the cold water and the other in the hot water. Hold them there for 90 seconds. Then place both simultaneously into the bowl of tepid water.

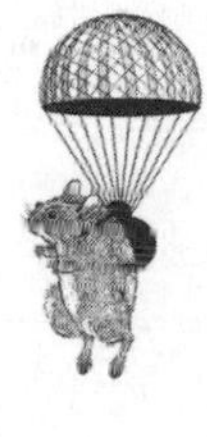

What will I feel? When your hands are respectively in the cold water bowl and the hot water bowl, one feels cold, the other hot. Hardly surprising. Yet, when you place them in the tepid water they still feel different, even though they are both now in water of the same temperature. The hand that was originally in the hot water now feels cold, but the hand that was originally in the cold water feels warm. After both have been held in the tepid water for a while the sensations ease off and eventually they feel the same.

What's going on? Human senses are relative, measuring only the differences between things that we perceive. So, your cold hand registers that the tepid water is warmer, and conversely your hot hand perceives the tepid water as colder. Your senses do not make absolute judgements on what is

around them, only the relative differences of their environments.

While this may lead to a few disconcerting moments, as your hands have just experienced, it helps us to focus on what is important and what is changing in the environment around us, and allows us to ignore things which aren't. It's a process called adaptation and is not just a product of our sense of touch. It affects all our senses and helps to prevent them suffering from overload. Without it we would find the world a debilitating place.

Adaptation is the reason we get used to smells and stop noticing them, or why our eyes become used to brightness or dark after we have been exposed to them for a little while. People who work in foul-smelling environments, such as refuse collecting or in fish-processing plants, are aware of the odour when they first arrive at work, but after a few minutes the impact is reduced as their sense of smell adapts to their surroundings. Similarly, people who work in noisy environments are able to filter out the constant background noise, allowing them to hear what their colleagues are saying once they have adapted to the other, permanent sounds.

PS: Adaptation may be the reason why, as we age, we believe that in our youth the sky was bluer, summer days were warmer and food tasted better. This was studied in the 1950s when researchers in the USA sat down volunteers in front of a screen and shone coloured lights onto it for 5 seconds. After a lapse of 5 more seconds, the volunteers were asked to adjust the controls on a coloured light generator to try to reproduce the colour they had just seen. The colours they chose were always brighter or deeper than those they had originally been shown. This is because adaptation was acting on their visual sense. They were given longer than the 5 seconds that they had viewed the original colour to set up their matching colour. So the longer they looked at the screen as they attempted to set up their match, the more the colour appeared to fade as their sensitivity to it declined. This meant that they continually turned up the brightness or depth of the colour as they worked at matching it to the original. The researchers believed that, because adaptation probably acts in the long term as well as the short term, the summer skies of our childhood seem brighter, the days warmer and the roast chicken more succulent. Other researchers dismiss this idea, saying ageing faculties are to blame. However, either argument might explain why Brussels sprouts, a particularly distasteful vegetable in childhood, never taste as bad in adulthood... although some colleagues in the *New Scientist* office dispute this heretical suggestion, insisting that they are as disgusting as they ever were.

Aromatic pee

Why does urine smell peculiar after you've eaten asparagus?

This really is quite a surprising effect, the onset of which can occur within minutes of eating asparagus. Once you recognise the smell it's impossible to visit a public lavatory without knowing if someone has been chomping on this expensive seasonal veg.

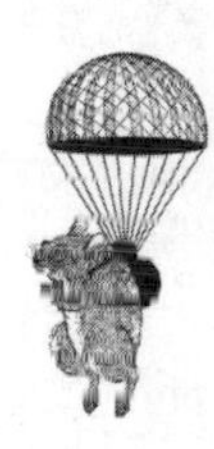

What do I need?

- cooked asparagus (boiled, salted asparagus is fine, but the author recommends his favourite recipe, which adds immeasurably to the pleasure of this experiment)
- a human urinary system
- a toilet

What do I do? Cook the asparagus to your preferred recipe, eat it and wait until you need to urinate.

What will I see? Not much, but you may be aware, on urinating, of a peculiarly pungent odour.

What's going on? The cause of the odour has perplexed biologists for a long time. While sulphurous compounds are known to be implicated, recent studies seem to suggest that it is in fact a cocktail effect, which combines to produce the unmistakable whiff. Methanethiol, dimethyl disulphide and dimethyl sulphone are the likely candidates. These substances are probably the result of the body breaking down the S-methylmethionine and asparagusic acid found in asparagus.

The difficulty in determining exactly which substances are responsible is compounded by the many variations in human production and perception of the odour. Several studies have indicated that production of the odour is a genetically determined trait exhibited by a maximum of 50 per cent of adults.

However, at least one report suggests that while all people produce the odour, only a minority are genetically able to smell it. So it may be that different individuals produce a different array of compounds and also have a differential sense of smell to these.

The reason that urine begins to smell so quickly after consuming asparagus may be due to the rapid response of the kidneys to any unusual products, such as the sulphurous compounds mentioned above. The body will quickly process and remove anything it considers potentially alien and the asparagus odour can often be detected in urine within 15 minutes of consuming the vegetable. Not that anybody should be alarmed – asparagus is an excellent, healthy food, it's just that your body removes the compounds it has no use for as quickly as possible while simultaneously absorbing the beneficial ones.

PS: Not everyone finds the smell offensive. Juvenal Urbino, in Gabriel García Márquez's *Love in the Time of Cholera*, 'enjoyed the immediate pleasure of smelling a secret garden in his urine that had been purified by lukewarm asparagus'.

The author's recipe for grilled, honeyed asparagus
Parboil a bunch of asparagus spears for 5 minutes – preferably the young tender tips. Meanwhile combine two tablespoons of honey with one tablespoon of extra virgin olive oil and mix well. Place the parboiled asparagus under a hot grill and cook until the spears just begin to char. Brush them with the honey/oil mixture and grill until the honey begins to caramelise. Serve hot, sprinkled with sea salt. If you are feeling particularly sophisticated, combine eating the asparagus with quaffing the champagne from the 'Fizz fallacy' experiment on page 28.

Wine into water

No matter what colour a drink is, when the liquid finally leaves your body the colour has gone. What happens to it?

Here's another one for the children. Ply them with cola, cranberry juice or orange squash and then set them thinking about why the ultimate product is always yellow.

What do I need?
- a variety of drinks of different colours: red wine (for the adults), cola, water, cranberry or orange juice, coffee or anything else with an interesting hue that you can think of
- a human digestive and urinary system
- a toilet

What do I do? Sample the drinks in turn after an initial visit to the toilet to ensure that the next visit's urine will contain mainly the products of your most recent drink. You may have to do this over a period of days in order to establish a rhythm, perhaps consuming each drink at the same time on consecutive days. Make sure that you don't have anything else to drink between consuming each liquid and visiting the toilet. Note the colour of your urine after each drink. If you are feeling adventurous after trying out the drinks from the list above, read below to see what happens if you eat too much borscht.

What will I see? You may notice variations in the yellow colour of your urine, but you will not, under normal circumstances, notice any colouring from the original drink showing through.

What's going on? Coloured substances in drinks (or food) are usually organic compounds that the human body has an amazing ability to metabolise, turning them into colourless carbon dioxide, water and urea. The toughest stuff is taken care of by the liver, which is a veritable living waste incinerator, while the kidneys take care of removing waste products from your blood. By the time it leaves your body the liquid is almost unrelated in chemical composition to the original liquid you consumed.

Any substance, solid or liquid, that goes down your oesophagus and passes through the digestive tract, if not absorbed, is incorporated into faecal matter. Urine, by contrast, is produced by the kidneys from metabolic waste produced in the tissues that has been transported through the bloodstream. This waste is added to any excess water you have consumed to produce urine of various shades of yellow. This is stored in and passed out from the bladder.

Any coloured compound that you drink either will or will not interact biochemically with the body's systems. If it does, this interaction (like any other chemical reaction it might undergo) will tend to alter or eliminate its colour. If it does not, the digestive system will usually decline to absorb it and it will be excreted in the faeces which, you will have noticed, show considerably more variation in colour than urine.

Some coloured substances can make it through your digestive system and into your urine. This occurs when the intake of coloured substances exceeds what the body can quickly metabolise and the colouring is not removed as the liquid leaves the body. Anybody who wants to see this effect should consume a large quantity of borscht (beetroot soup). You'll notice that your urine takes on a distinctly pinkish hue. (If you are one of those rare people with highly acidic stomach conditions, then you may find the pigment is broken down more rapidly. Nonetheless, if you look carefully enough, there should still be a slight pinkish change in the colour of your urine.)

PS: There is a current fad for drinking water in order to purify ('detox') the system after an excess of anything from alcohol to red meat to Christmas dinner. But evidence suggests this is unnecessary. The minimum volume of urine required by the kidneys to excrete the waste products of the human metabolism is about half a litre a day. Because we lose another half litre through breathing, sweating and defecating we need to replace the total lost by drinking about 1 litre a day. Drinking more than this merely dilutes the urine – the same amount of waste products (or 'toxins' as they are popularly known) are still removed from our bodies. Coffee and tea have a mild diuretic effect, but your body still gets a net fluid gain from drinking them – you do not urinate more liquid than you consume. Alcohol, on the other hand, is a much stronger diuretic, so you should always drink lots of soft drinks or water after over-indulging. 'Detoxification' may be popular, but there is no evidence to support the argument that our kidneys cannot cope with the input and output of normal or even excessive consumption as long as the latter is not on a long-term basis. Drinking more water than we need makes no significant difference to the elimination of waste products from our bodies.

A long meal

So now we know what can and can't pass through our bladders, what about our bowels? How long does it take to digest different types of food?

Children will love this one – the gross factor looms large. And for the really brave, there can be fun with a kebab skewer afterwards!

What do I need?

- a human digestive system
- sugar
- streaky bacon
- sweetcorn
- tomato
- mushrooms
- celery
- beetroot
- a stick or kebab skewer for poking through the products

What do I do? Cook and eat the foodstuffs, plus anything else you fancy – fibre-rich vegetables produce the most interesting results.

What will I see? Possibly nothing, depending on how closely you are prepared to inspect your faeces. Plant material such as red tomato is often visible in human excrement, as are the skins of peppers and capsicums. Sweetcorn is probably the most obvious retained product of the human digestive system, its yellow colour standing out particularly well in faeces, although celery and mushrooms are also difficult to break down, while beetroot can give a fascinating reddish hue, as it does to your urine.

What's going on? Simple sugars such as glucose will be absorbed into the bloodstream relatively quickly and will not appear in any noticeable form after passing through the body. They may even pass directly from the mouth into cheeks and gums because their molecules are small enough to pass directly into human cells.

However, most food needs processing by our digestive system before it will pass into our bloodstream and become useful fuel. The journey begins in our mouth where the food is crushed until the pieces are small enough to swallow, and enzymes begin to break it down further. Once in the stomach, it is attacked by more enzymes and acids in gastric juices and is blended by the churning motion of the muscles in the stomach wall. A couple of hours later this semi-liquid passes into the small intestine, from where much of it is absorbed into the bloodstream.

Some components, such as dietary fibre, cannot be digested and move on to the large intestine. This takes about six hours. Once there, more water is absorbed by the body so that only the indigestible material remains. This may take up to 36 hours before the remaining waste is expelled through the anus. The final faecal form is approximately 75 per cent water. The rest is bacteria, excreted biliary compounds and unfermented fibre.

Of the various foodstuffs we eat, sugar is the first to be absorbed, followed by proteins (from eggs, nuts and non-fatty meat) in about six hours, and finally by various types of fat over longer periods, although the age, health and size of the eater will affect all these timings.

As you will discover from carrying out the experiment, some components of food, such as fibre, are hardly broken down at all and, as in the case of sweetcorn, pass out of the body relatively untouched. The lignins in mushrooms and the cellulose in celery are also relatively unscathed after a journey through your body.

Those adults brave enough – and you can guarantee children will be keen – can, conditions and consistency willing, take a long kebab skewer and poke through the poo to see what they can recognise. If you do feel up to attempting this, disposable gloves are a sensible precaution, and you should certainly wash your hands thoroughly after disposing of the experimental material.

PS: *New Scientist* readers are an ingenious bunch. In answer to a related question on how much excrement a human produces for a given amount of food eaten they came up with the following information.

How much excrement you produce depends not just on how much food you have eaten, but the type of food and the activity of the bowel. If you eat lots of high-fibre foods such as vegetables, beans and cereals which, as we have seen above, the body does not completely digest and absorb, you will produce more faeces than if you eat lots of easily digested, low-fibre foods such as chocolate. Spicy foods, laxatives and infections can affect the activity of the bowel. And the greater the speed of transit, the less water the gut can absorb and the greater the weight of faeces produced, as anybody who has had gastroenteritis will testify.

The weight of faeces produced by a healthy individual ranges from 19 to 280 g per day. The only way to increase faecal weight is to eat more fibre, because unfermented fibre is indigestible and can hold lots of water. In healthy people, the wet faecal weight is in the order of 3–5 g/1 g fibre. From this we are delighted to be able to extrapolate the following formula for the effect of fibre in the colon:

$$\text{Faeces weight} = W_f(1 + H_f) + W_b(1 + H_b) + W_m(1 + H_m)$$

where W_f, W_b and W_m are respectively the dry weights of fibre remaining after fermentation in the colon, bacteria present in the faeces, and osmotically active metabolites and other substances in the colon which could reduce the amount of free water absorbed. H_f, H_b and H_m are constants denoting their respective water-holding capacities.

Want to read more?
www.constipationadvice.co.uk/constipation/digestive_system.html

Wayward water

Why does an electrostatically charged plastic comb attract a stream of water?

Don't be concerned by the terminology. To charge your comb electrostatically all you need to do is to comb your hair with it...

What do I need?

- a plastic comb (not a metal one)
- some hair
- running water
- black treacle

What do I do? Comb your hair, or someone else's if, like the author, you don't have any. Turn on your tap until you get a thin, smooth, steady flow of water. The thinner the stream, the better, as long as it does not break down into droplets. Then place the comb near the point where the water emerges from the tap. Afterwards, try the same experiment with black treacle. (And take a look at the 'Shape shifting' experiment on page 62.)

What will I see? The water is attracted towards the side of the flow where you are holding the comb. Large deflections of the water stream can occur.

Comb your hair

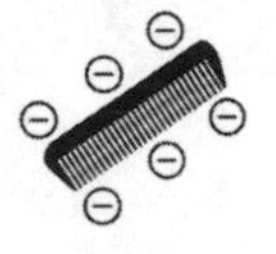

Comb is now negatively charged with electrons

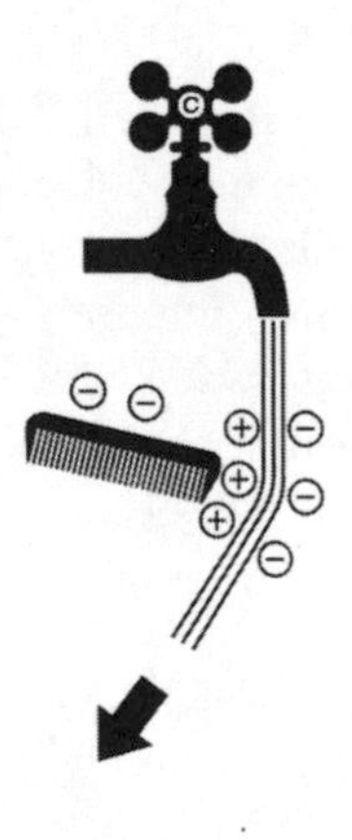

Hold the comb near a slow, thin flow of water from a tap.

The electrons in the comb attract positive charges in the water molecules, (and push away negative charges).

What's going on? The molecules of water from the tap are electrically neutral overall but, like the molecules of other dielectric substances, they are characterised by regions of positive and negative charge. The electrical field caused by the comb (which is produced by an excess of electrons collected by the comb from the friction caused as it runs through your hair) attracts the more positive parts of each water molecule and repels the negatively charged regions.

The water molecules are rotated and stretched by the field, and tend to become aligned head-to-tail along the electrical field lines, with their positive ends pointing towards the comb. The charges on each molecule therefore neutralise each other except at the water/air interface that is closest to the comb, and also at the interface furthest from it. In these regions they are not paired with opposite charges (because they are at the edge of the water stream and there are no molecules on one side of them) so they build up areas of equal positive and negative charge on the opposing sides of the cylindrical water surface. The stream is now polarised and it is this property that allows dielectric materials to be attracted to a source of electrical field – in this case the comb.

Water is not unique in showing what is called the Doff effect. Try the same experiment with black treacle. Dribble a thin filament of treacle from a tablespoon. Because the treacle is more viscous than water and therefore flows more slowly, the electrical field from the comb has more time to act on each section of the flow's length. The resulting deflection is spectacular.

PS: A similar principle applies when you rub a party balloon on your clothing and then stick it to the wall. The negative electrons collected by the balloon repel electrons in the wall's surface, leaving behind a net positive charge that attracts the balloon, which will stick there as if suspended in air.

Spectral images

When you draw pictures in condensation on a bathroom mirror, they disappear when the condensation evaporates, but return when the room steams up again. Why?

This is a perfect one for the children, but if you'd prefer them not to write rude words on your mirror which will reappear when Great-aunt Gladys takes a bath, challenge them to a game of noughts and crosses. If you win, you can revisit your victory the next time you take a shower.

What do I need?

- a bathroom that readily steams up
- a mirror
- a finger

What do I do? Wait until you are taking a hot bath or shower (*New Scientist* is very green, so there's no excuse for water wastage or excess heating) and your bathroom has steamed up, then draw or write on your mirror with your finger. After you've finished bathing, open the windows and let the room cool. The next time you or a family member takes a bath and the room steams up, look at the mirror.

What will I see? Your original message or picture, which vanished when the room cooled down and the condensation on the mirror evaporated, will return when the bathroom steams up and the mirror mists over.

What's going on? When water vapour condenses on a dry mirror, it does so as separate droplets – a process known as dropwise condensation. These droplets effectively screen the mirror so that it appears opaque.

When you draw on the surface with your finger, the droplets coalesce into a thin film of transparent water, so the mirror becomes reflective again in these areas – this is called filmwise condensation. When the mirror warms up, or the air humidity falls, the droplets evaporate and the image disappears because the surrounding droplets no longer contrast with it.

However, the film of water evaporates more slowly than the droplets because of its lower surface area. If it does not have time to evaporate completely before the bathroom steams up again, any condensation occurring soon afterwards will be dropwise where there were droplets before, and filmwise where some of the film remains. The image will then reappear on the glass.

However, this only explains how mirror drawings return in the short term. If the mirror dries completely, the pattern should not normally reappear when further condensation occurs. But this presumes that your drawing finger is entirely clean, which is unlikely. When you draw an image in the condensation your finger will almost certainly leave behind traces of grease or sweat, plus possibly shampoo or soap. These traces are transparent, so when the condensation disappears you can't see them. The next time water vapour condenses on the mirror, however, there is a noticeable difference in droplet size between those forming on the clean glass and those forming on the greasy or soap-contaminated glass.

Grease will tend to repel water droplets, while water-loving surfactants such as soap will reduce the droplet size and generate a smoother, clear film of water. Whatever the cause, as the mirror steams up again, the image you drew will contrast with the opaque mist on the surrounding glass.

PS: Dropwise condensation is known by chemical engineers to be more efficient at transferring heat than filmwise condensation, but in practice it is much more difficult to promote, because as the droplets enlarge, they touch each other and coalesce, so the process tends to become filmwise.

On the other hand, dropwise condensation is easy to prevent. Wiping the mirror with a cloth or a tissue wetted with a small amount of detergent such as shampoo leaves an invisible film on the surface, as we have seen. This reduces the surface tension of the condensing droplets, causing them to flatten out and readily coalesce into a film. This is the basis of anti-misting fluids used for treating spectacle lenses and the inside of car windscreens.

Brush mush

Why does orange juice taste so awful after brushing your teeth?

You know how it is. You got up late and you're in a rush. You swig a glass of orange juice right after you've brushed your teeth. Instead of a sweet fruit flavour, the juice tastes like a foul bitter concoction. At its worst, the rest of your breakfast can taste revolting too. Something's going on…

What do I need?

- orange juice
- three types of toothpaste: 1. ordinary mint; 2. mint without foaming agents (this is available at wholefood shops and you should check the label for the absence of sodium lauryl sulphate [SLS], which is a foaming agent used in many toothpastes); 3. non-mint (again, one that doesn't contain SLS)

If you can't get hold of the SLS-free varieties of toothpaste, you can still try the experiment using the ordinary mint toothpaste.

What do I do? Take a swig of orange juice and note how sweet it tastes. Brush your teeth with non-mint toothpaste, then take another swig of juice and see how it tastes. Repeat with toothpaste number 2 and then toothpaste number 1.

What will I taste? After brushing with non-mint, non-foaming toothpaste, you might expect the juice to taste foul, but surprisingly, it tastes just like normal sweet juice. Brushing with non-foaming mint toothpaste makes the juice taste weird, but not too bad. However, normal mint toothpaste transforms the juice into a bitter, medicine-like flavour and strips all the sweetness away.

What's going on? You've just experienced the 'orange juice effect', something that's been known about since the 1970s. Foaming agents, usually the sodium lauryl sulphate, are added to toothpaste to help disperse the paste around your mouth, and for easy rinsing, but they have a nasty after-effect. They interfere with the taste buds on your tongue, suppressing the ability to taste sweetness and salt, and enhancing any bitter flavours. The powerful mint flavour that most of us want from our toothpaste simply adds to the problem. While it doesn't interfere with our taste buds in the same way as the foaming agents, the strong mint flavour overpowers the taste of anything that is consumed afterwards, including the breakfast you were looking forward to. The rule – and it makes sense for the health of your teeth too – is brush your teeth after meals, not before.

PS: Foaming agents aren't a vital ingredient, but, as with soap, if we don't see any bubbles we think it isn't working. The same goes for mint. If a toothpaste doesn't knock you out with its mint flavour, we assume it's not effective. In Japan, most toothpastes are fruit-flavoured, and in the West many children's toothpastes are fruity, not minty, too.

5 In the Garage

(or the utility room, workshop or garden shed...)

Hot stuff

Does hot water really freeze faster than cold water?

Astoundingly, yes it does. And not only that but *New Scientist* magazine can quite smugly lay claim to being the first to publish this discovery in the modern era when a Tanzanian student, Erasto Mpemba, wrote to the magazine in 1969, pointing out that when he was trying to make ice cream he found that the mixture froze more quickly when put in the freezer warm than if allowed to cool to room temperature first. Readers were sceptical, but experiments with both ice cream and water – distilled or tap – proved that Erasto had spotted a quite extraordinary phenomenon. But why does it happen?

What do I need?

- flattish plastic containers or ice-cube trays
- warm water at about 35 °C
- cold water at about 5 °C
- a freezer

Or:

- two metal buckets
- warm water at about 35 °C

- cold water at about 5 °C
- freezing night air

You'll also need a thermometer to measure the water temperature if you want to be truly accurate.

What do I do? Fill one container or tray with the warm water, a second with the cool water and place them in the freezer. Give them 10 minutes at least before checking them, then check again at regular intervals. If you are using the buckets and a cold night, you'll need to fill them and wait some time longer.

What will I see? The warm water will freeze faster than the cold water. It really will! The only limitation in the case of the freezer experiment is that the water containers must be relatively small so the capacity of the freezer to conduct away the heat content is not a limiting factor.

What's going on? Initially it was believed that placing the containers on a freezer shelf allowed the container with the warmer water to melt the icy surface on which it was resting so that it settled into a nice frosty resting place. That meant more of the container was surrounded by ice and therefore it had greater contact with cold surface than the cooler water container. The increased rate of heat transfer from the warmer container and its contents therefore more than offset the greater amount of heat that had to be removed.

However, while this is clearly an accelerating factor, further research showed that the effect occurred even if the containers were sitting on a dry surface or were suspended above the freezer shelf.

It was then suggested that a reduction in the volume of the warmer water due to evaporation might make that body of water smaller and so cool more quickly, but this is not the case. Thermometers placed in the water show that the cooler water reaches freezing point more quickly than the warmer water, as you would imagine and as predicted by Newton's law of cooling. After that, however, the water that had started warmer solidified more quickly. In fact, in experiments, the maximum time taken for water to solidify in the freezer occurred with an initial temperature of 5 °C and the shortest time at about 35 °C (which is why we suggest those temperatures when describing how to carry out the experiment above).

This seems paradoxical, but can be explained by understanding that there is a vertical temperature gradient in the water. The rate of heat loss from the upper surface is proportional to the temperature of the surface. So, if the surface can be kept at a higher temperature than the bulk of the remaining liquid (in this case, the initially warmer water), then the rate of heat loss will be greater than from liquid with the same average temperature uniformly distributed (in this case the initially cooler water). This can be proved by using tall, thin metal cans as water containers. The paradoxical effect disappears in this case because the temperature gradients in the cans are lost as heat is rapidly conducted through the metal sides much faster.

Nonetheless, there may be other factors at work. Cold water forms its first ice as a floating skin which impedes further convective heat transfer to the surface. Hot water, on the other hand, forms ice over the sides and bottom of the container, while the surface remains liquid and relatively hot, allowing radiant heat loss to continue at a faster rate. The large temperature difference drives a vigorous convective circulation which continues to pump heat to the surface even after most of the water has become frozen.

This argument is further borne out when conducting the experiment using metal buckets full of water in open air on a cold, preferably windy, night (which promotes heat loss). If the initial temperature of the water is around 5 °C, cooling of the core is very slow, particularly as loose ice floats to the surface, inhibiting normal convection. There is no means by which the slightly warmer water in the 5 °C bucket can come into contact with the cold bucket edges and transfer its energy to the outside.

But if the initial temperature of the water in the bucket is around 35 °C, strong convection is established before any water freezes, and the entire mass cools rapidly and uniformly. Even though the first ice will form later than in the cool water bucket, complete solidification of the water in the warmer bucket can occur more quickly than if the water starts off cool.

But those temperature differences are critical. Obviously, if the cold bucket starts at 1 °C and the hot at 99 °C the experiment is unlikely to cause surprise. Additionally, the containers must be large enough to sustain convection with a small temperature gradient, but small enough to extract heat quickly from their surfaces. This is where the forced cooling of a windy night helps.

And there's more: recent, as yet uncorroborated, research from Washington University in St Louis, Missouri, has offered yet another possibility. Solutes, such as calcium and magnesium bicarbonate, precipitate out if water is heated. These can be seen lining the inside of any kettle used to boil hard water. However, water that has not been heated still contains these solutes and as it freezes the ice crystals that are forming expel the solutes into the surrounding water. As their concentration increases in the water that has yet to solidify they lower its freezing point in the same way salt sprinkled on a road in winter does. This water therefore has to cool further before it freezes. Additionally, because the lowering of the freezing point reduces the temperature difference between the liquid and its surroundings, the heat loss from the water is far less rapid. Nonetheless, the fact that both tap water and distilled water (the latter being virtually free of solutes) both display the effect described in our experiment shows that this is not the decisive factor.

Finally, there could be yet another factor affecting this process – supercooling. Research shows that because water may freeze at a variety of temperatures, hot water may begin freezing before it is cold. But whether it will completely freeze first may be a different matter.

Just one more issue remains, but this is a technological, rather than a physical dimension. The temperature oscillation inside a freezer depends on the sensitivity of the thermo-element and the timer that controls the system. We may assume that at the freezer's standard temperature the power used for cooling the freezer operates at a standard rate. So if a bucket of cold water is added, it may produce only a small effect on this power output because it will not trigger the temperature sensor. However, a bucket of hot water may easily activate the sensor and release a short but powerful cooling of the freezer with an additional cooling overshoot, depending on the timer. While this would not account for the version of the experiment carried out outdoors on a cold night, it might explain some of the freezing differences seen in the ice cube/domestic freezer version of the trial. This would only apply, however, if the experiment was being carried out with the hot and cold containers being placed in the freezer independently of each other in two separate experiments. By placing them in the freezer together, as suggested above, the variable can be excluded.

It seems that a number of factors are at work here, with no one theory acting in isolation. Do take time to try out any number of combinations of temperature, vessels and environments – after all, fame awaits whoever stumbles on the ultimate answer.

PS: This effect was noticed by Sir Francis Bacon in 1620 using wooden pails of water standing on ice. Even earlier, in 300 BC, Aristotle also noted that hot water froze faster than cool water. This is Aristotle's account from *Meteorology*: 'Many people, when they want to cool water quickly begin by putting it in the sun. So the inhabitants when they encamp on the ice to fish (they cut a hole in the ice and then fish) pour warm water round their rods that it may freeze the quicker; for they use ice like lead to fix the rods.'

Nonetheless, and despite the phenomenon's illustrious history, it was Erasto Mpemba who dragged the paradox into the modern era. If anybody knows what became of him or where he lives now, we'd be delighted to hear from him.

Freezer teaser

How can you grow spikes of ice in your freezer?

If you are tired of mini-umbrellas and olives on cocktail sticks in your drinks, cubes of ice with dangerous-looking spikes sticking out of them might just enliven your parties.

What do I need?

- a domestic freezer (preferably a frost-free model)
- two plastic ice-cube trays
- tap water
- distilled water

What do I do? Fill one tray with tap water, the other with distilled water and place both in the freezer. If your freezer has a temperature control it seems –7 °C is the ideal spike-growing environment.

What will I see? Ice spikes should begin to grow from the surface of the ice-cube tray containing distilled water. They are less likely to grow out of the control tray containing tap water, but they might. Sometimes no spikes will form in either tray so repeat the experiment until you see what you are looking for.

What's going on? First, a bit of history. Ice spikes have been noted growing out of frozen bodies of water for some time, but nobody knew why they only formed occasionally. Then, in 2003, Ken Libbrecht and Kevin Lui from the physics department at the California Institute of Technology in Pasadena began to look into the mystery.

Libbrecht had been growing designer snowflakes in his laboratory to find out why they formed such complex and delicate patterns and was used to creating specialist ice formations. When he was sent photographs of spikes protruding from a tray of ice cubes he was intrigued and started trying to grow them himself in his home freezer. Not surprisingly, most

Fill the tray with water and place it in the freezer. Ice forms first at the edges and on the surface, leaving a hole in the middle.

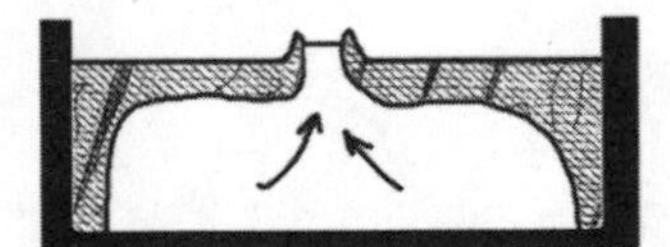

The ice thickens and pushes water up through the hole where it freezes.

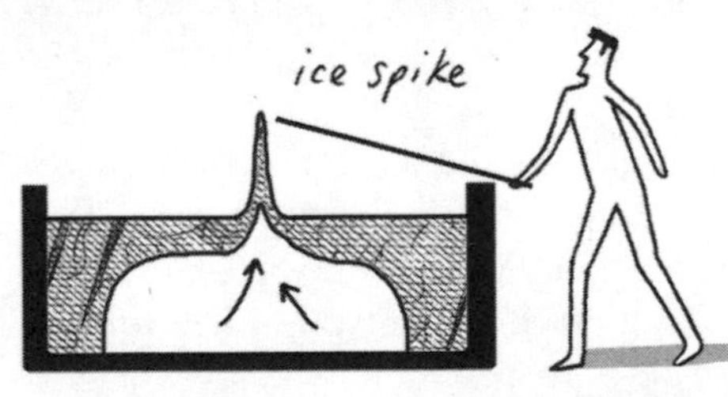

Eventually the tip freezes over creating the ice spike.

of the cubes he froze turned out – well – cubic, but once in a while a spike would grow out of one of them. When he began trying the experiment with different compositions of water he discovered that distilled water produced more spikes per ice-cube tray than tap water.

Why? After about 90 minutes in the freezer the water surface on the ice cubes starts to freeze, initially around the edges of the cube. This is because the plastic tray is covered with microscopic nicks and scratches and water molecules can wedge themselves into these tiny hollows. Here the molecules can form plenty of bonds with their neighbours as the temperature falls without being moved around in any convection currents occurring in the bigger body of water. Because ice crystals are less dense than water, they float to the surface. The freezing ice then creeps slowly towards the middle of the water surface until only a small hole remains unfrozen at the centre of the ice cap.

At the same time more ice starts forming around the sides and bottom of the cube. Because ice expands as it freezes, the ice growing below the surface pushes water up through the hole. If the conditions are just right, the meniscus of water forced out of the hole freezes around its rim, forming the base of a hollow spike. As this process continues, the spike grows taller until all the water has travelled up its hollow interior and frozen or, more commonly, the tip of the tube freezes over. Interestingly, spikes reach their full height in no more than 10 minutes.

The reason that tap water produces fewer spikes than distilled water is due to the impurities in the former. As the tap water freezes around the top of a growing tube of ice its saltiness increases because any dissolved minerals and metals remain in solution as long as unfrozen water is present and they are therefore contained in the last bit of the water to freeze. They build up to such high levels at the tip of the spike that the water there can no longer freeze and the growth

stalls. So to get the best ice spikes use distilled water – the kind sold for car batteries or steam irons.

Libbrecht and Lui have more advice for those wishing to grow ice spikes. Temperature makes a big difference. At –7 °C, about half of distilled water ice cubes turn spiky, but only 1 in 10 show a spike at –15 °C. This may be because the tips of the embryonic ice spikes freeze over before they have a chance to grow. And a frost-free freezer helps because in these models a fan is continually circulating cold, dry air. The moving air chills the edges of the water droplet perched atop a growing ice spike faster than in calm conditions and this promotes the growth of longer ice spikes. This works even better if you allow the air to circulate around your freezer by removing all those pizza boxes and frozen chicken wings. Finally, plastic ice-cube trays work far better than metal ones. Metal conducts heat away from the cubes so quickly that the water surface freezes over too quickly, sealing the hole from which the spike would normally grow.

PS: It's possible to grow ice spikes in your garden, and these can be even more spectacular than those grown in your freezer. You need a small container such as a plastic saucer or bowl which you should fill to just below the rim – again you'll have better results with distilled water. You'll also have to wait for a freezing night which, as in the case of the frost-free freezer, would be made more ideal by the presence of a light breeze. If you are fortunate, you'll get an ice tower taller than those seen in the freezer and, if the breeze is light and constant but not strong, the tower will probably lean in one direction. It may even be triangular in cross-section as the wind flows around the forming tower. A garden ice spike grown in New Zealand in July 1991 formed a helical icicle with eight complete turns as the breeze changed direction in swirling conditions.

It seems the record length for a freezer spike grown from an ice-cube tray is 5.6 cm but those grown from double-distilled water in larger containers can reach 10 cm. Ice spikes grown in gardens can exceed 25 cm in height. If you find a foolproof way of beating these records, we'd love to hear from you.

Want to read more? The 20/27 December 2003–3 January 2004 issue of *New Scientist* (p. 38) has full details on the discovery and creation of ice spikes.

Cubed route

Why is the highest concentration of bubbles in ice cubes always found in the middle of the cube?

The answer to this question is closely related to the formation of ice spikes we saw in the 'Freezer teaser' experiment immediately before this one. To prove (a) that scientists are a lazy bunch and won't do any work beyond that which is necessary and (b) why the bubbles form in the pattern noted above, you can prepare this experiment at the same time as growing your ice spikes and run both alongside each other.

What do I need?

- a domestic freezer
- ice-cube trays
- tap water

What do I do? Fill the tray with tap water and place it in the freezer. If you are running the 'Freezer teaser' ice-spike trial alongside this one, you'll have to set the freezer temperature control (if you have one) at –7 °C, although it's not necessary for the purposes of this particular experiment.

What will I see? When the cubes are frozen you'll notice that in most cases there is a concentration of bubbles in the centre of the cube, with spines of progressively smaller bubbles radiating outwards.

What's going on? Two effects covering the solidification of ice are at work here: the segregation of a solution into two separate physical forms or phases as it solidifies; and the way in which ice crystals tend to grow.

Tap water contains on average 0.003 per cent dissolved air by weight, but the solubility of air in ice is very small so the first crystals that start to grow after you have placed your ice-cube tray in the freezer contain almost no dissolved air. As the tray cools, heat flows away from the water through each of the outer faces of the cube and the first crystals start to form on the outer walls of the cubes, growing inwards as more ice appears in the form of columnar crystals. These early, outer crystals are clear, because they contain no dissolved air.

However, the unfrozen water which still lies between these early, airless crystals and in the centre of the cube gradually become saturated with the air that is being driven out of those first crystals. When the concentration of air in this remaining water reaches 0.0038 per cent it forms a peritectic composition – a mixture that has the lowest freezing point of all possible mixtures of that particular substance. When this occurs all the remaining liquid freezes, separating into a mixture of approximately 2.92 per cent air by volume, and 97.08 per cent ice by volume.

The smaller bubbles radiating outwards that you see in the frozen cubes are from the peritectic mix which was trapped between the early, clear columnar crystals, and the high concentration of bubbles in the middle is where the remaining bulk of the peritectic mix had been trapped. Hence the traditional ice-cube bubble pattern.

Now pop the cubes you've just made into the James Bond vodka martini left over from the 'You only drink twice' experiment on page 11 and sit back knowing more about the science of enjoying yourself than you did before.

PS: An extreme example of the formation of columnar crystals can be seen if you bite into an ice lolly. You'll see crystals radiating inwards from the outside of the lolly towards the centre. These were created as they grew inwards from the mould the lolly was made in. You might also see columnar crystals and the segregation of dissolved elements (in the same way the bubbles form in ice cubes) in cast metals and candles.

Incidentally, it is very difficult to create bubble-free ice cubes; the ones you see in TV ads are usually fakes, made from glass or perspex so they won't melt under the studio lights. However, if you want to try to make some at home it can be done. Use boiling water to remove some of the dissolved gases. Leave the boiled water to cool in its container covered with cling film to prevent air from re-dissolving in the water. Now freeze the water very slowly so that there is no steep temperature gradient and the bubbles, when they appear, do not get trapped but have time to move through the liquid and up to the surface before freezing into place.

Commercial ice-cube machines create clear cubes by freezing a small portion of water at a time then trickling more over its surface so that the gas-carrying component runs away while layer after layer of non-gassy ice builds up.

Cleaning up

Why is cola so good for cleaning dirty coins?

It's not only dirty money that you can launder with cola; brass and similar alloys can be brought to sparkling lustre using your favourite fizzy drink.

What do I need?

- a handful of loose change (preferably copper coins)
- a few cans of cola
- a large glass or bowl

If you are happy with the results of cola on your coins you might also like to get hold of brass or other alloy household objects in need of a clean, such as doorknobs or screws.

What do I do? Fill the glass or bowl with cola and drop in your coins.

What will I see? Leave the coins for a few hours while you try out one or two of the other experiments in this book. When you return you'll notice how bright and shiny your once tarnished coins are. If they are particularly dirty you may need to give them a rub with a tissue or cloth, but when you do you'll be pleased to see they look as though they were minted yesterday.

What's going on? The tarnish that forms on an alloy is caused by a reaction between oxygen and one of the metals that make up the alloy. This forms an oxide. Cola, which is weakly acidic, reacts with the oxide to form a soluble salt of the metal that produced the tarnish, which then dissolves in the liquid, leaving behind a clean metal surface as bright and shiny as the day the coin was produced.

Cola contains phosphoric acid. Because this acid can also eat away at substances other than the tarnish on coins, you should always brush your teeth after drinking cola. And because regular cola and diet cola both have the same cleaning effect, don't be fooled into thinking that diet cola may in some way be better for your teeth than regular cola. It does have less sugar, but the phosphoric acid levels are very similar.

The citric acid in lemons and the acetic acid in vinegar are also excellent for removing the oxide on copper coins. If you clean a lot of copper coins in vinegar you'll notice it takes on a green hue, which is the copper acetate formed during the reaction between copper oxide and acetic acid. Beware though: copper salts are poisonous, so don't consume the vinegar (or, indeed, the cola) afterwards. And remember, while this method is great for cleaning your loose change, it should not be used for valuable collections of antique coins and the like.

PS: *New Scientist* reader Rob Hay wrote in to assure us that even in the light of the effect that cola has on coins, it has little effect on the stomach – despite its acidity – because the stomach is even more acidic than the cola. In fact, he has anecdotal evidence to show this. As a keen and sometimes thirsty cyclist in his youth he'd often spilt soft drinks over the paintwork on his bike with no ill effects. But on one cycling trip he became rather ill and, as he was dismounting from his bike, he was violently sick. Ninety minutes later the areas of the frame that had come into contact with his vomit had been completely stripped of paintwork. The contents of the stomach, it seems, are far more corrosive than the contents of a can of cola.

Sticky solution

How does molasses clean rusty objects?

Molasses rivals cola in its ability to buff up tarnished household objects. Which foodstuff you prefer to use probably depends on what it is that needs cleaning.

What do I need?

- molasses
- water
- a large plastic bucket
- some rusty household items (door hinges, bicycle chains, old nuts and bolts all fit the bill)

What do I do? Mix one part molasses to nine parts water in the plastic bucket. Completely submerge the rusty item in the liquid and leave it for two weeks.

What will I see? The article will emerge clean and shiny from the mixture.

What's going on? Molasses contains chelating agents. These are made of molecules shaped a bit like the claws of a crab – the word chelation comes from the Latin *chele*, meaning claw. These agents can envelop metal atoms on the surface of an object, trapping them and removing them. Molasses owes its properties to cyclic hydroxamic acids, which are powerful chelators of iron.

More of these compounds are found if the molasses is derived from sugar beet rather than cane sugar. The plants from which molasses is made presumably use these chelating agents to extract minerals from the soil. Interestingly, there are aerobic microorganisms that use similar cyclic hydroxamic acids to scavenge iron. So plants and microbes appear to use the same chelation strategy to obtain their daily ration of iron.

A similar process is at work when old coins are cleaned with cola (see 'Cleaning up' immediately before this). Phosphoric acid, a key ingredient of cola, is a chelating agent. The power of a chelating agent also explains why the insides of tomato cans are lacquered. The citric acid in the tomatoes would dissolve the metal of the can if the lacquer were not present. Household cleaning agents, especially detergents and shampoos, also rely on chelation. These soften water to make it more effective during the cleaning process.

Chelation has its uses in medicine too. EDTA (ethylenediamine tetra-acetic acid) is used as a chelating agent to control calcium in the body and can reduce the effects of mercury or lead poisoning.

PS: Brown sugar is the wholemeal flour of the sugar world and white sugar is the refined version. When white sugar is extracted from the juice of sugar cane, the dark, non-sugar substances left behind are known as molasses. Most of this is used as animal feed, but molasses is also popular with health food consumers and vegetarians as it is an excellent source of minerals and vitamins – a tablespoon of blackstrap molasses has an iron content similar to that of about 300 g of sirloin steak, though it is still 70 per cent sugar.

Sticking doors

Why does a freezer door that has just been opened and closed stick shut so tightly that it is almost impossible to reopen?

We've probably all noticed this. You take out the ice cream and shut the door only to realise that you've forgotten to take out tomorrow's chilli con carne. But the door is now resolutely stuck…

What do I need?

- a freezer that has been switched on for at least 24 hours

What do I do? Open the freezer door for a few seconds, then close it for a few seconds. Now try opening it again.

What will I see? The second time you try to open the door it's much more difficult. In some cases you'll end up pulling the freezer towards you as you tug at the handle. In others, the door will just remain closed and you'll have to wait before the effect wears off and you can open it again as easily as the first time.

What's going on? Freezer doors stick because opening the door the first time allows some of the cold air to flow out of the bottom of the freezer. If you stand in bare feet at the door of an open freezer you will notice this cold, dense air chilling your toes. The cold air flowing out at the bottom allows warmer air at room temperature into the top of the freezer.

When the freezer is closed again, this new air cools and contracts, creating a partial vacuum and making the door seem to stick as it is pressed to the seal by the higher air pressure outside the freezer.

If the day is particularly warm, there is a greater temperature difference between the exterior room temperature and the interior freezer temperature. In this case it is sometimes necessary to prise open the door seal a little with your fingers to allow some air in to equalise the pressure inside and outside the freezer before you can get the door open.

The partial vacuum created soon passes, however, because the door seals are not absolutely airtight and the air pressure inside and outside the freezer is equalised by air leaking into the compartment.

It's worth pointing out that if you try this experiment and you don't notice the sticking-door effect, you should check the seal. It may be leaking a lot of cold air and consequently making your freezer work much harder than it should do, to say nothing of the environmental and monetary costs involved.

PS: The vacuum effect is one reason among many why it is very dangerous to let children play with fridges and freezers. Certainly they should be told that they should never climb inside an empty one. This experiment should be carried out only when an adult is present.

6 In the Garden... and Further Afield

Fossil record

When he dies, I'd like my pet hamster to become a fossil. How can I make sure this happens?

Remember, whatever applies to a hamster applies to humans too. So if you would like to become a fossil, pay attention to what follows, make detailed notes in your will and find a sympathetic funeral director.

What do I need?
- a naturally deceased hamster (or other pet)
- a variety of environmental conditions

What do I do? Take your hamster to one of the natural environments described below and ensure that it will not be removed from its final resting place by scavengers or natural phenomena.

What will I see? Very little. Fossils take tens of thousands of years to form, but you will be saving up enjoyment for future generations of palaeontologists.

What is (should be) going on? A desire to preserve our fluffy pets' remains in fossilized form may be admirable, but the fact that they have a hard, mineralised skeleton and live a non-marine lifestyle is a bad start. Terrestrial conditions are liable to erosion and significantly reduce the chance of fossilization. The soft tissue surrounding the skeleton of mammals decays very quickly – it is usually preserved only if the animal dies at high altitude or in a freezing environment such as a glacial crevasse or in the polar regions, and then only in wizened, mummified form, which is not true fossilization.

So, if you really want your pet to survive the ravages of geological time, then while it is still alive you need to concentrate on improving the quality of its teeth and bones. Fossilization of these involves additional mineralisation, but you can give your hamster a head start by feeding it a diet rich in calcium to build up its bones and teeth. If you are seriously thinking of becoming a fossil yourself, you should get a good dentist. If your hamster has eaten a few seeds in its last days, these too can become fossilized and intrigue palaeontologists in the millennia to come.

After that it's a matter of location, location, location. You need to bury your dear departed rodent where it won't be disturbed for a very, very long time. Fossilized remains are often found in caves, so you could take up potholing to scout out suitable locations (after proper training, of course). Alternatively, you need to find a spot where your hamster will be buried quickly after it has been laid to rest – preferably somewhere natural and dramatic, the sort of site from which you can detect a distant volcanic rumble and clouds of ash being emitted skywards. Don't get too close though: an ash burial is good, incineration by flowing lava is not. Again, be very careful, you don't want to join your hamster in fossilized harmony at this stage.

As the volcanic option suggests, you may have to travel a long distance to find the optimal conditions. A desert wadi in the flash-flood season offers a good environment, as does a tropical river floodplain during heavy rain so you can bury your hamster in fine, anoxic (oxygen-free) mud.

However, the best environment to ensure a fossilized hamster is probably a sea burial in very deep water (shallow marine conditions are turbulent and full of life which will disturb or eat the remains). There are few creatures in the deep sea, and even fewer below the sediment and mud. As long as your hamster is not buried near a tectonic subduction zone where the Earth's crust is being consumed, such as the Pacific coast of the Americas, it should rest undisturbed until fossilization takes place.

This environment and the land-bound ones described above are ideal. The oxygen-free conditions will slow decay of the body and the ash or fine seabed clay will help to preserve the body structure. Fossilization will then proceed until your hamster is nothing more than an outline of carbon and petrified body fluids, thanks to compaction from the weight of sediment that settles above. You should allow at least 200,000 years for this.

Of course, there is more chance of winning the lottery than you or your pet ending up as a fossil, but that's no reason not to try.

PS: The following advice to let stalactites do the work of an embalmer is found in this poem written by Richard Whatley in 1820. The subject of the poem, William Buckland (1784–1856), was one of the most famous geologists of his day – and a noted eccentric, who claimed to have eaten his way through the entire animal kingdom. His contemporary, Augustus Hare, recorded how Buckland came upon a casket containing the preserved heart of a French king. Buckland exclaimed, 'I have eaten many strange things, but I have never eaten the heart of a king before,' and promptly gobbled it up, the precious relic being lost for ever.

'Elegy Intended for Professor Buckland'

Where shall we our great Professor inter,
That in peace may rest his bones?
If we hew him a rocky sepulchre,
He'll rise and break the stones,
And examine each stratum that lies around –
For he is quite in his element underground.
If with mattock and spade his body we lay
In the common alluvial soil,
He'll start up and snatch those tools away,
Of his own geological toil;
In a stratum so young the Professor disdains
That embedded should lie his organic remains.
Then exposed to the drip of some case-hardening spring,
His carcase let stalactite cover,
And to Oxford the petrified sage let us bring,
When he is incrusted all over;
There, 'mid mammoths and crocodiles, high on a shelf,
Let him stand as a monument raised to himself.

Finally, and it should go without saying, if you intend to attempt fossilization of your pet, wait until it has died of natural causes before carrying out the above instructions.

Goggle-eyed

Why can I see clearly underwater when I'm wearing swimming goggles or a mask, but everything looks blurred if I don't?

Our eyes are adapted to viewing in air, but what goes wrong when you open them underwater without first putting on your goggles?

If, like the author, you wear contact lenses, make sure you remove them before attempting the part of this experiment that requires you to open your goggle-free eyes underwater.

What do I need?

- a large volume of water (a swimming pool – or a sun-kissed ocean surrounding a tropical island will show the effect at its best)
- swimming goggles or a diving mask
- your eyes

What do I do? Don the goggles or mask, take a deep breath and place your face in the water and look around. Hold your hand at arm's length in front of your face and check its clarity. Now come up for air! Take the goggles off and do the same again.

What will I see? If your goggles don't mist up, you'll be able to see objects underwater clearly, including your fingers. After you've removed the goggles you'll notice that everything you look at is blurred.

What's going on? Light travels more slowly through water than it does through air. When light moves from one medium to another and undergoes a change in speed, the beam is bent, or refracted. The amount of refraction depends on the ratio of the speed of light through each medium. You can see this effect when you place a spoon in a glass of water. Where the spoon meets the water it appears to bend – this is refraction.

The human eye is delicately balanced to make sure that light entering the pupil is focused by the lens onto the retina at the back of the eye. But this system is optimised for light that has passed through air before reaching the eye. The eye has evolved to take account of the refraction that occurs when the light strikes the air/eye interface and not a water/eye interface.

When light passes straight from water to the eye, it is bent by a different amount, and so isn't correctly focused by the lens. Goggles restore the air/eye interface and normal sight is resumed.

Glasses and contact lenses take advantage of light refraction in different media by bending the light that passes through them before it reaches the eye, correcting imperfect vision in short- or long-sighted people.

PS: For those of you who prefer the answer in mathematical form, here goes…

The amount of refraction of the light depends on the difference between the refractive indices of the media on the two sides of the surface of the eye's cornea (in one case the media are air/cornea; in the other they are water/cornea). The bigger the difference, the more light is bent. Because the refractive indices of air, water and the cornea are 1, 1.33 and 1.38 respectively, this difference is much smaller when the eye is in contact with water than when it is in contact with air.

The refractive power P of a surface is given by:

$$P = (n(1) - n(2))/R$$

where $n(1)$ and $n(2)$ are the refractive indices of the cornea and the medium outside it respectively and R is the radius of curvature of the cornea. P is measured in dioptres. A dioptre is the unit of refractive power, equal to the reciprocal of the focal length (in metres) of a given lens. Assuming a value of 0.008 m for R, the refractive power in air is about 47 dioptres and in water is about 6 dioptres.

We can vary the focusing ability of our eyes to a degree. However, the increase which this can produce is much less than the loss of 41 dioptres when the eye is in contact with water. In fact, the biggest change the eye can achieve is about 15 dioptres.

Because the eye cannot refocus to compensate for the lost 41 dioptres when it is immersed in water, objects viewed without goggles will always seem blurred.

Head trauma

How much does a human head weigh?

The author had the dubious pleasure of carrying out this experiment on live television. It has certainly been one of the most popular questions we have answered down the years in *New Scientist*, and here's the lowdown on how you can carry out the experiment yourself.

Because it involves immersing your whole head in water, make sure that there are always at least two people present and that children are supervised.

What do I need?

- your head
- a bucket of water filled to the brim
- a larger, empty vessel big enough to accommodate the bucket and its contents

If you don't have a larger vessel, you'll need a measuring jug.

What do I do? Place the full bucket inside the larger, empty vessel. Take a deep breath and lower your head, crown downwards, into the full bucket until the water reaches the base of your chin. Hold your head still for as long as possible until any water ripples cease, then take it out and have a few gulps of air. Because a lot of water is splashed about while carrying out this experiment, unless you have a very large bathroom, it is probably best to wait for a warm, sunny day and then go outside.

What will I see? Water will be displaced by your head, spill over the sides of the full bucket and be collected in the larger vessel. Collect all the water that has spilt into this outer vessel and measure its volume. If you don't have a larger vessel to catch the water, let all the displaced water run away and then start to refill the bucket into which your head was placed with a measuring jug taking a note of exactly how much water you need to refill the bucket to the brim.

What's going on? Measuring the weight of your head involves effectively isolating it from the rest of your body. But while decapitation is an obvious advantage it has far too many drawbacks, not least that you will not be able to see the results of your experiment.

Fortunately, there are a number of solutions. The first is the least practical. Your neck vertebrae are responsible for holding your head's weight. If you hang upside down the vertebrae in your neck move apart slightly because of the weight of your head pulling on them. So to weigh your head you simply lower yourself slowly onto a scale while hanging upside down. While you are doing this you need to continually observe the distance between the top vertebra of your neck and your skull, using, say, an ultrasound scanner. The instant the top vertebra starts moving towards the skull stop lowering and read the scales. Because your neck is not imparting any force onto your head this isolates your head from your neck, thus giving you an accurate measure of your head's weight (other than a small amount via the elasticity of your vertebrae and the spring coefficient of the balance).

Of course, most people have neither the inclination to hang themselves upside down, nor have the use of an ultrasound scanner. Happily there is an easier way and it is surprisingly accurate.

It works because although bone and brain are constituent parts of our heads, the rest, like the remainder of our bodies, is mostly water. We know that 1 litre of water weighs 1 kg, so if we can measure the volume of water displaced by a head we can approximately work out how much that head weighs. If you displace 4 litres of water using the experiment above, your head weighs approximately 4 kg.

Tests on members of the author's family and colleagues suggest that the average weight of a human head is 4.25 kg.

PS: Readers should take care not to follow the lead of the author, who was conned into believing the experiment would only work in water at or near an excruciating 4 °C, the temperature at which 1 litre of water weighs exactly 1 kg. This is unnecessary because the volume of water displaced is the same whatever its temperature, even if its weight is less because it is warmer and less dense than it would be at 4 °C. We suggest using warm water unless you want to have blue ears and nose.

Complete collapse

Is it true you can crush a can just by heating up water inside it?

If you take great care, yes, you can, and you'll also see the awesome force that the weight of the atmosphere can exert on people and things that are on the surface of the Earth. Just stand back while you are doing it...

What do I need?

- water
- a large metal can with a screw top (the square type with a metal carrying handle used to package motor oil or olive oil is fine; a tin with a press-fit lid also works)
- a stove or gas-fired barbecue (we actually used a charcoal barbecue, but if you do the same, be prepared to wait)
- heat-resistant gloves (such as oven gloves)

For those who are wary of handling old petrol cans and flames you can try a version of this experiment using a plastic fizzy drinks bottle with a screw cap and some hot water.

What do I do? Head out to the garden or other open space. Wash out the can, making sure no residue remains. If it has contained motor oil you may need to use a strong detergent. Add enough water to cover the bottom 2 cm and very carefully stand the can upright over the flame with the top off. You'll probably want to put a child cordon around the area – this is very much light-the-blue-touch-paper-and-stand-well-back territory. When the water starts to boil vigorously, don the gloves, remove the can from the heat and screw the top firmly into position. Now wait a few seconds.

What will I see? The can will begin to bend and twist and buckle, often making a lot of metallic-rending noises. It's quite spectacular and you end up with a piece of crushed and twisted metal.

What's going on? This experiment shows the phenomenal weight of the atmosphere at the earth's surface. When you remove the can from the heat and replace the top tightly, the steam inside that has been rapidly expanded by the hot water begins to cool and contract. This means that the space inside the can is no longer at atmospheric pressure and can no longer support the sides of the can against normal atmospheric pressure outside. The result is the spectacular crushing of the can.

PS: The pressure of the atmosphere – caused by the weight of the air above – is said to be 1 atmosphere at sea level, or 101.325 kilopascals, or 14.7 lb per square inch, depending on the units you use. This is a huge figure – the weight of a column of air 1 m^2 in cross-section reaching from the Earth's surface to the top of the atmosphere would be the equivalent to a mass of 10.2 tonnes at the surface – yet it is one which our bodies bear easily. Its immense force, however, can be seen in practice when applied to objects like our low-pressure can.

For a milder version of this experiment you'll need an empty, flexible plastic bottle and some hot water. Carefully pour the hot water into the bottle, screw on the top and shake the bottle. Unscrew the bottle taking great care. High pressure inside can cause hot water to spray as the cap is removed, Empty out the water, screw the top back on and watch. After a while the bottle sides will begin to collapse inwards. Much the same has happened here as happened with the can. The hot water heats up the air inside the bottle, expanding it and forcing some of it out of the bottle. When the air cools the pressure in the bottle falls and, because the top is tightly fastened, no air can rush into the bottle to restore normal atmospheric pressure. The air pushing on the outside of the bottle therefore has higher pressure and the bottle is crushed.

Mixing madness

Is it true that if you mix the right amount of corn flour with the right amount of water you end up with a liquid that solidifies on impact?

You'd better believe it because knowing this could save your life! A liquid that solidifies on impact sounds like a contradiction in terms, but that's exactly what you get if you mix corn flour and water in the right quantities. Such liquids are known as dilatant materials and are utterly spooky because the harder you stir them the more solid they become. Stop stirring and you have a liquid again – and playing with these materials is positively mind-bending. What's more, should you ever find yourself stuck in quicksand, you'll be glad you read this book.

What do I need?

- 300 g corn flour
- 250 ml water

- a metal bowl
- a spoon
- a hammer

You will also find a saucepan useful if you want to take your experimenting even further.

What do I do? Mix the corn flour and water in a medium-sized bowl until it becomes too hard to move the spoon. Stop stirring and tip the bowl. Then stir again vigorously. You can also run your fingers through the mixture or even try to strike it with a hammer.

What will I see? After you have mixed the liquid with the spoon and it has become stiff, stop stirring. Now tip the bowl and you'll notice the mixture becomes a liquid again and flows. If you stir it again it immediately thickens once more. Try dipping a finger or the spoon into the mixture very slowly – if you are extremely gentle it will remain liquid, but pull out your finger or the spoon in a hurry and it will solidify. Drag your fingers quickly through the mixture and you'll be able to lift out a putty-like ball that you can work in your hands. But if you open your fingers and stop roiling it, it will quickly become a liquid again, so keep your hands over the bowl. Now try striking it with a hammer (this explains why you need a metal bowl). If you time the blow just right, the mixture will shatter. Best of all, the broken pieces will re-liquefy and pool together like the shape-shifting creatures in *Star Trek: Deep Space Nine* or *Terminator* 2. You can even throw the material against an outside wall and watch it shatter. Bear in mind, though, that while this is fun, it's also very messy.

What's going on? The mixture you have created is what is known as a dilatant material, or a shear-thickening fluid (STF). In these materials viscosity increases as the force, or shear, on them is increased. So, as you have already discovered, the more pressure you apply, the more resistant to deformation they become. The more effort you put into stirring the mixture, the more it will resist. This is because the application of force to the material causes it to adopt a more ordered structure. Under normal conditions the particulates in the liquid are only loosely arranged, but the shock of any impact or pressure changes their alignment, locking them into place. When the stress dissipates the material relaxes again.

Mining companies have to be aware of the dilatant properties of fluids. High-solid slurries such as mixtures of water and particles of mined kaolin or calcium carbonate being pumped away from mines and quarries have to be carefully monitored. If the quantities of water and particles hit critical limits the pipelines can rapidly fill with solid dilatants, blocking the entire system.

Research into STFs has led to the development of 'smart materials', ones that respond to changes in their surrounding environment. For example, military researchers are attempting to treat fabrics with STFs so that when a bullet strikes an STF treated uniform, the uniform becomes rigid at the point of impact and the bullet fails to penetrate it. Under normal conditions, however, the fabric would be as flexible as normal clothing.

However, corn flour is made up of particles with a very high proportion of starch, which gives it another fascinating property – when heated it can be converted from an STF into a thixotropic fluid, a fluid that becomes more runny when stress is applied – in other words, almost the opposite of a dilatant material. (Thixotropic fluids are discussed in the 'Saucy stuff' experiment on page 39.) Corn flour's transformation from a dilatant to a thixotropic fluid is achieved by diluting the thick paste with more water and heating it in a saucepan. The starch particles in the corn flour break open to release long molecules of starch. As we saw in 'Saucy stuff', this leads to thixotropy. Corn flour is a quite incredible household substance.

PS: Thick slurries of beach sand are often dilatant, leading to that old favourite of film directors: the victim trapped in quicksand. As the person struggles the sand takes a firmer grip. But any student of dilatant materials and now every reader of this book knows how to save themselves and defy Hollywood legend. Gentle, sedate movements (like a slow breaststroke), not kicking legs and arms, provide the route to safety.

Tea-souper

Why does soup cool more quickly than tea or coffee in a vacuum flask?

If you're trekking in the Alps or cross-country skiing in Arctic climes, this is a serious consideration. You don't want to open your flask to find its contents don't have the warming effect you'd banked on. So just why do watery beverages stay hotter longer than viscous ones?

What do I need?

- two identical vacuum flasks
- a pan of hot, thick soup (vegetable soup containing lots of puréed solid matter works best)
- freshly made coffee or tea

What do I do? Pour the soup into one flask, the tea or coffee into the other, and seal both. Take a sip from each flask after an hour and note which of the two liquids is hotter. Repeat this at hourly intervals.

What will I see? You'll find that tea or coffee stays hotter for longer than the thick soup.

What's going on? The link between heat (or thermal energy in transfer) and temperature is one of the trickiest concepts that students encounter when studying physics. Soup in a vacuum flask may start at the same temperature as the tea or coffee, but the soup's temperature falls faster for the following reasons.

When thermal energy (measured in joules) is transferred from the soup to the flask, the temperature of the soup falls according to the following equation:

Thermal energy transferred = mass of soup (in kg) × specific heat capacity of soup (joules per kg per °C) × change in temperature of soup (°C)

The important factor here is the specific heat capacity of the soup, which is the number of joules that are needed to change the temperature of 1 kg of soup by 1 °C. Water has a very large specific heat capacity: 4,200 joules per kg per °C. The specific heat capacities of tea and coffee are slightly lower, but fats (and the solid matter in the viscous soup) have much lower values and hence their temperatures drop far more for the same thermal energy flow through the flask. Such foods also heat up and cool faster in all situations. A good example is full-fat milk, which boils much more rapidly than water or, indeed, skimmed milk.

Water has this especially high specific heat capacity thanks to the unique characteristics of the hydrogen bond. Astoundingly, you need to extract more energy from 1 kg of water than from 1 kg of any other common material to cool it by 1 °C. This means that wise outdoor types know that only water should be carried in vacuum flasks – all additives such as tea, coffee and freeze-dried food are taken in dry form and added later. This means that although your soup will be less tasty than if it was made fresh at home, it will be hotter when you consume it – a vital consideration if you are experiencing extreme conditions.

Interestingly, despite its being an important constituent of many vegetable soups, the potato's high water content makes it an ideal thermal energy reservoir – hence the jacket potato, so useful for warming hands on Bonfire Night. And beware

French onion soup. Its very watery base means that its specific heat capacity approaches that of water but it is also protected by a floating, anti-evaporation film of oil. This can blister the tongue of the unwary diner while companions are already eating their third helping of viscous bouillabaisse. The good news in both cases is that if you're thinking of making soup before you set out for the day, potato or French onion soup should be your choice if you want to eat it hot.

One more confounding effect remains. If you are too lazy to stir your soup properly while it is heating, it may boil in only a few local spots before it all reaches 100 °C, rather than having an even heat spread throughout. If you put the soup into your vacuum flask at this stage it will start off at a lower temperature than boiled tea or coffee and so become tepid even sooner.

PS: We were delighted to hear from the Harrogate Ladies' College, who provided us with evidence to help us formulate our studies in this field. Under controlled conditions they measured the cooling curves, boiling points and the specific heat of both tea and mushroom soup. Their graphs showed that soup does cool faster than tea and their recorded specific heat figures of 2960 joules per kg per °C for soup versus 4185 for water easily account for this effect. Remember – real scientists experiment!

Repulsive vegetables

Why do cabbage leaves repel water so well?

Use discarded leaves from the kitchen chopping board to check out which are waterproof and which are not.

What do I need?

- cabbage leaves

- leaves from other plants and vegetables such as lettuce, cauliflower, spinach, honeysuckle and dock
- a supply of water – a watering can will do; heavy rainfall is better

What do I do? Sprinkle the leaves with water. If you know it's going to rain, line them up outside and watch how they repel or collect water.

What will I see? The cabbage leaves and some others, like honeysuckle, will repel the water drops, forming beads that run off the surfaces. If you tip the leaves they will become almost completely clear of water as it readily pours off the surface. Other leaves, such as lettuce, will allow the water to collect on their surfaces and, given enough coverage, will actually be damaged by the quantity and weight. Left out in overnight rain, cabbage leaves will have only a few drops on their surfaces, while lettuce leaves will be covered in a layer of water.

What's going on? Most leaf types are coated in water-repellent waxes. However, cabbage leaves (and the leaves of other plants including honeysuckle and nasturtium) have a super-hydrophobic surface, one with an extra layer of powdery wax forming tiny pillars. When a droplet of water lands on such a surface it touches only the tips of the wax pillars and not the leaf itself. This causes the water to form beads which easily run away on anything other than a totally flat surface. Even if they sit on the surface they can be removed simply by tipping the leaf.

This does not happen on leaves that are not super-hydrophobic. On these leaves surface tension flattens out the water droplets and allows them to stay in their resting place.

You can remove the waxy surface from the cabbage by rubbing it. The leaf then acts like a lettuce leaf. You will also notice that the cabbage leaf seems greener where you have rubbed it because the wax scatters the light that hits it, giving cabbages their bluish-white sheen.

The leaves of such plants are often close to the ground, where their surfaces pick up dust, dirt and bacteria, so when water rolls off their surfaces, they are effectively cleaning themselves. However, their surfaces reflect more sunlight than other leaf types so this might explain why not all plants use this system, because it reduces their ability to photosynthesise.

Sound bites

Can I measure the speed of sound?

You can indeed, although the experiment is a bit low-tech compared with the 'Hot chocolate' experiment on page 79, used to measure the speed of light. You'll also need access to a very large space…

What do I need?

- a hammer
- a helper capable of safely wielding the hammer
- a hammer-resistant and preferably resonant surface (a wall, piece of metal or rock will do)
- a stopwatch (or a clock with a second hand)
- a very long measuring tape or a measured ball of string
- a pair of binoculars
- a very large garden or flat open space such as a beach or a park

Add a metal coat hanger, some string and a spoon to check out further properties of sound.

What do I do? It goes without saying that this job must be carried out by an adult as a hammer is used.

Ask your helper to begin striking the hard object once every second with the hammer – they'll need to use the stop-watch and maintain a regular beat. Begin walking away from your helper, looking back from time to time. When you are a few hundred metres apart, use the binoculars if necessary.

What will I see? As the distance between you and your helper increases, the delay between them striking the hard object and the sound reaching your ears will become greater. Eventually, the delay will match the time between each beat and the sound will once more appear to coincide with the action of the hammer.

What's going on? Sound has a relatively low speed through air, which means you can measure it if you have an open space large enough. Sound travels at 344 m per second in dry air at 20 °C. This may vary at different temperatures, but not by much, so this experiment works pretty well just about anywhere.

When the sound of the hammer once again coincides with the 1 second beat, you are ready to measure the speed of sound. Stop walking away at this point and measure the distance between you and the helper. You should be 344 m away, or very close to that.

Of course, very few of us have such long gardens, and finding a flat, open space of that length, unless you live near a beach, may prove troublesome. Then there will be the problem of outside noises interfering with you detecting the beat of the hammer. If this is the case, ask your helper to increase the frequency of the beats to once every half second (in which case you should find yourself about 172 m from your helper when the beats again coincide with the sound), or once every quarter second if they can (when you'll be about 86 m away).

PS: The speed of sound is slow enough for noises to be noticeably delayed when heard from even short distances, as this experiment verifies. You've probably already experienced this when seeing a car door being slammed when you are 100 m or so away, or even when you have heard your own voice echoing back after shouting in a tunnel or beneath a bridge.

However, in liquids and solids, the speed of sound is noticeably increased. This is because sound transmission is propagated through neighbouring atoms or molecules. The speed of sound depends on the interactions between these atoms and molecules – in effect, how often they collide. In air this is relatively low, but in seawater the speed of sound is 1500 m per second while in steel it reaches 5100 m per second.

In order to check out just how sound travels more efficiently through solids and liquids try the following. Take a triangular metal coat hanger and tie a piece of string to each of the two corners that don't carry the hook. Then tie one of the strings to your left index finger and one to your right. Now lift the coat hanger so it is dangling, hook down, from the two strings. Ask a friend to tap the coat hanger with a spoon. You'll hear a tingling sound much as you'd expect. Now, still holding the strings, put your fingers in your ears and get your friend to tap the coat hanger again. This time the sound will be more reminiscent of a peal of bells and much, much louder. That's because the sound is travelling through the solids and liquids of your body rather than through the air.

Ba(n)g

Why does mixing vinegar with baking soda produce an explosive mixture?

It's amazing how useful vinegar and baking soda are to the amateur scientist. Stand back and admire their versatility...

What do I need?

- warm water
- vinegar
- a measuring jug or coffee mug
- baking soda
- a tablespoon
- a tissue or small muslin bag
- a clear plastic food bag
- a tight-fitting plastic kitchen clip (or two if you have deft hands)

You'll also need to make sure the bag doesn't leak, so take a big breath and blow it to check that it isn't deflating from a minute scratch or flaw.

What do I do? Take everything outside – the results can be quite messy. Place two tablespoons of the baking soda inside the tissue (or muslin bag) and fold the tissue around the powder so you have a neat packet. Put 150 ml of vinegar (about half a mug) and 75 ml of warm water (about a quarter of a mug) into the bag and seal it with the clip. Take the baking soda package, open the bag, drop the package in and seal the bag as quickly as possible with the clip. To make sure it doesn't leak you can add a second clip but be quick, the bag can inflate rapidly. Then shake the bag. Drop it and step back quickly.

What will I see? The bag will expand, then explode, often quite loudly and always messily.

What's going on? Baking soda is essentially sodium bicarbonate and this reacts with the acetic acid in vinegar to produce carbon dioxide, which takes up much more space than the powder and liquid from which it is formed. Therefore, the gas expands quickly to the point where it can no longer be contained by the sealed bag. Hence the explosion.

PS: You do need to be careful when you are attempting this unless you want to find yourself covered in smelly liquid. The reaction is very rapid, so instead of shaking the bag you can just drop it and run after putting the baking soda package inside.

Fizzy flight

Alka-Seltzer tablets can be used to make film canisters fly through the air. How?

If, like the author, you've hung on to your 35 mm SLR camera you'll be well set to try this experiment. If you haven't, you'll have to get hold of some old 35 mm film canisters. It's worth the hunt – or worth buying a film – in order to see this splendid effect. The flying canisters move at considerable speed, so make sure everybody is standing well back.

What do I need?

- empty 35 mm film canisters
- Alka-Seltzer tablets
- water
- outdoor space

What do I do? Half-fill the film canister with water. Pop in the Alka-Seltzer tablet and quickly snap on the lid as firmly as possible. Shake the canister and stand it upside down in the middle of your garden and step back quickly.

What will I see? The canister will pop and fly upwards leaving the lid behind on the ground. Tests in the *New Scientist* back garden saw the canister achieve heights of around 2 m.

What's going on? This miniature launch station takes advantage of the simple reaction that occurs when Alka-Seltzer is added to water. Alka-Seltzer is an antacid designed to combat the unpleasant effects of acid indigestion in the stomach. For this reason it contains sodium bicarbonate, an alkali (hence the 'Alka' part of the brand name) that neutralises the acid. It also contains citric acid which, when added to water, reacts vigorously with the sodium bicarbonate to form carbon dioxide. This reaction is similar to the one seen in the 'Ba(n)g' bag experiment immediately before this.

The carbon dioxide bubbles form quickly inside the film canister, causing the pressure to rise until it is so great that the canister is blown apart between lid and base.

PS: Children love this experiment. While you need to make sure they don't get their hands on the tablets or get struck by flying canisters, they can design coloured pieces of card which can be folded into tubes and stuck round the canisters. Add a cone to the top of the tube and fins around the base and you can hold a competition to see whose rocket flies highest and furthest.

Blow out

Try out two different types of volcano in your garden…

Not all volcanoes look like Mount Vesuvius – they come in different shapes and sizes and emit different products. You can try out two of the most common on a miniature scale. This is not as spectacular as the 'Overreaction' experiment involving cola and Mentos (see page 194). Even so, it's best done outside.

What do I need?

- two bottles of fizzy drink – the bigger the better (2-litre bottles work well), and coloured drinks such as Tizer, Irn-Bru or cherryade give a pleasing effect
- wallpaper paste powder
- a funnel
- a fridge
- outdoor open space

What do I do? Chill the bottles of fizzy drink to a temperature as low as possible without freezing the liquid. Take the bottles out of the fridge, open one without shaking it too much and pour out enough to make room for the wallpaper paste powder you'll be adding to it. Put the funnel into the open neck and add a few tablespoons of wallpaper paste powder until the bottle is full (you can experiment with different amounts of paste and drink). Quickly screw the cap back on. Now let both bottles warm to ambient temperature (a warm day is better for this experiment than mid-winter). Shake the bottles vigorously, place them in the centre of the open space about a metre apart, and quickly unscrew their caps.

What will I see? The bottle that does not contain the wallpaper paste powder will spray its liquid out in a wet spray, the gas bubbling out through the liquid with very little of the drink leaving the bottle before it dies down again. The liquid that does come out will spread out in a puddle.

The bottle that contains the wallpaper paste powder, however, will produce a thick stream of gunk, oozing quickly from the bottle, building up around the base and spreading out much more slowly than the liquid around the other bottle.

What's going on? First, a bit of background. You chill the drinks to ensure that the bubble-making carbon dioxide that fizzy drinks contain remains in solution – the colder the drink, the more carbon dioxide it can hold. Carbon dioxide also stays in solution when under pressure, so while the caps of the bottles remain in place, the carbon dioxide remains dissolved in the liquid. When the drinks warm up before you unscrew the caps to carry out the experiment, the carbon dioxide becomes less soluble. So, when the pressure is released as the caps are removed, it rapidly begins to bubble out of the drinks, forcing the liquid out of the bottles.

Once the bottles are opened you are, in effect, creating two types of volcano in garden-scale form. The bottle containing the wallpaper paste resembles a volcano emitting felsic lava. This is highly viscous and contains more than 63 per cent silica. As the lava bubbles up inside the volcano it decompresses, just like the fizzy drink does as it leaves the once pressurised bottle, and gas forms in the viscous fluid. Because the fluid is sticky in both felsic lava and the wallpaper paste drink, the bubbles cannot escape easily so the volume of liquid swells and is forced upwards out of the bottle. Consequently, lava in this form tends to be more explosive than other types as the gas struggles to escape the confines of the volcanic vent and the viscous liquid and can blast its way out. This explosive action and the thick lava give rise to the classic, cone-shaped volcano we associate with Hollywood movies.

Genuine examples include Mount Vesuvius, in Italy, and Mount St Helens, in the USA.

The bottle without the wallpaper paste powder resembles a volcano emitting mafic (or basaltic) lava – so called because of its high magnesium and iron content. This lava has less than 52 per cent silica and is far runnier. Because of this, the gas is able to bubble free after the lava decompresses at the Earth's surface. This allows the liquid rock – like the pasteless fizzy drink – to run smoothly from its vent with little or no explosive effect. This leads to long flows of molten rock and volcanoes with shallower slopes because of the more fluid lava, accounting for those seen in Iceland and Hawaii. Nonetheless, while the volcano slopes are shallow, they can grow to great heights because they are rarely destroyed in catastrophic eruptions. In fact, the largest lava shield – as these volcanoes are called – is found in Hawaii. Known as Mauna Loa, it is 120 km in diameter and rises 9000 m from the sea floor to form part of Hawaii's Big Island. Even that is dwarfed by Olympus Mons on Mars, a shield volcano creating the tallest mountain in the solar system at 27,000 m.

PS: Many types of wallpaper paste contain a fungicide to stop mould growing on interior walls. Don't be tempted to drink the volcano once it has finished erupting!

Up and down

How does a yo-yo yo-yo?

Indeed, how does a yo-yo do lots of different things? When you've mastered the technique of making the yo-yo travel up and down its string and read the how and why, have a go at a few of the special tricks we list at the end.

What do I need?

- a yo-yo
- a bit of practice

What do I do? Play with the yo-yo in the normal way, making it travel up and down its string. If you've never done this before, it's quite tricky at first, so if you are a novice here are a few tips to get your experimentation under way, thanks to Learn2Yo-Yo (www.tutorials.com/09/0915/0915.asp). Slip the middle finger of your hand through the loop, placing the loop in the centre of your finger, between your first and second knuckle joints. Let the yo-yo dangle for a while to unravel any kinks in the string. Then wind the yo-yo up its string and cup it in your palm with your middle finger placed in the string groove and your other fingers curved around the sides. Bend your arm to a right angle at the waist, with your forearm in front of you and your elbow at your side. Lift your forearm slightly, snap your wrist back, then drop your arm down to its original spot and release the yo-yo as if you're throwing it to the floor. Your hand should open up completely at this point. As the yo-yo travels down the string, lift your forearm and hand slightly in a smooth motion. This gives it extra speed. You'll feel a slight jerk of tension when it reaches the bottom. The yo-yo should then start its journey back up. As it does, drop your forearm and hand in the same easy motion, adding a bit more kick. The yo-yo should return to your hand just as your arm reaches waist level (the same position you released it). You can catch it here and start again, or keep your hand open and continue the downward arc of your arm, so your palm hits the yo-yo and pushes it down for another pass.

What will I see? The yo-yo will travel down its string, reach the bottom and travel back up again in apparent defiance of gravity.

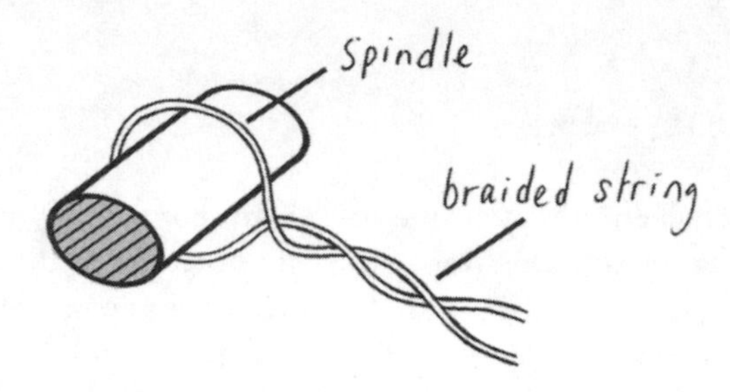

What's going on? If you examine a yo-yo, you'll see it consists of two heavy discs connected by a smooth axle. The string is not tied to the axle but simply looped around it. You can visualise this string as a 'U' which starts at your finger, goes down and around the yo-yo's axle and back up again to your finger. The string is twisted, forming a loop at the bottom so that the yo-yo can't jump out of the 'U'. The string is, in fact, the most important part of a correctly set-up yo-yo and should always be the single 'U' piece looped in half, and not tied to the axle as many often are.

As the yo-yo is released from your hand, its initial potential energy is converted into angular kinetic energy. It stores most of this energy, like a gyroscope, in its spin or angular momentum. The angular momentum of a rotating object is the measure of the extent to which the object will continue to rotate about its axis unless acted upon by an external force. A good yo-yo has more mass on the outside of the disc than on the inside in order to keep it light while at the same time maximising the angular momentum.

The end of the string loops loosely around the axle of the yo-yo, leaving the yo-yo free to spin inside the loop when it reaches the bottom of its descent. The friction between the string and the axle here is not great enough to allow the axle to grab the string and start winding itself up. However, if you give a small jerk to the string – this happens in normal yo-yo usage when the string jerks as the yo-yo reaches the bottom of its descent and is outlined in our guide to using a yo-yo above – this momentarily reduces the string's tension. The friction of the string against the axle is then enough to allow the string to start wrapping itself around the axle.

Once the string has made one turn around the axle, the loop starts to act as if it is attached – the outer turns of string stop the inner loop from slipping any more and the yo-yo then begins to climb back up the string, converting its kinetic energy into potential energy.

The propensity of the yo-yo to remain spinning depends on the friction between the axle and the string. This can be adjusted by twisting the string to make it tighter or looser around the axle, or by waxing the inner axle to lubricate it. The smoother and looser it is, the easier it is to do some of tricks we list below, but the harder it is to make the yo-yo climb back up the string.

PS: Now that you know how to use a yo-yo and understand how it works, you may feel confident enough to try out some of these special tricks.

'Sleeping' a yo-yo involves dropping the yo-yo down its string in the normal fashion, but not pulling it back up so that the yo-yo remains at the bottom of the string spinning freely. The key to sleeping the yo-yo is to have a loose loop around its spindle. A tight loop, as described earlier, causes the string to grab the spindle, making the yo-yo travel back up after it hits the bottom of its cycle. The loose loop allows the yo-yo to spin freely.

'Walking the dog' involves sleeping the yo-yo after giving it a particularly hard throw to put lots of energy into its spin and then allowing the spinning yo-yo to touch the ground. It will 'walk' along the ground spinning on its axle.

'Spanking the baby' is achieved by first sleeping the yo-yo and then slapping the back of the string-holding hand after a few seconds of sleep. This jars the string and, if the loop is not too loose, will cause momentarily increased friction and the

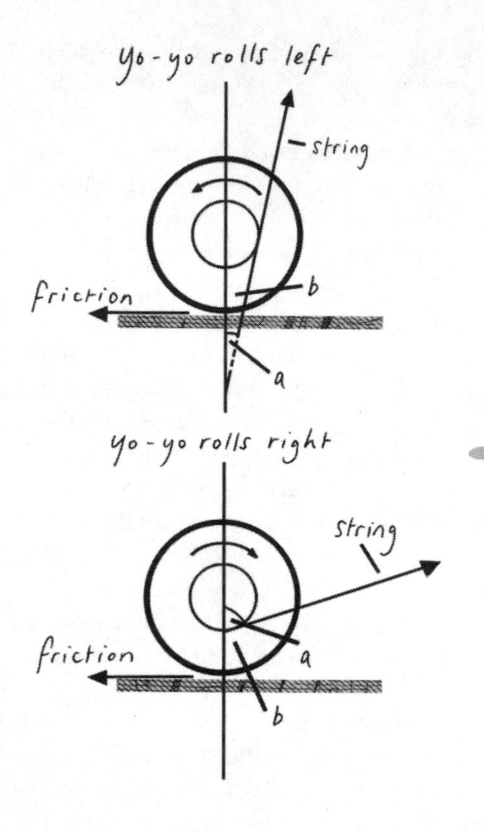

string will catch the spindle, returning the yo-yo back to your hand.

PPS: There's another property of a yo-yo which demonstrates how friction can act on bodies. Place your yo-yo on a horizontal surface and look at it side-on with the thread coming out of the bottom of the spindle to the right. If you pull the string to the right, which way will the yo-yo move? The answer is left, right, or not at all – it depends on the angle *a* at which you pull the string.

When *a* is small and the thread is pointing upwards, the thread tends to turn the yo-yo anti-clockwise. However, this is counteracted by friction: *b* is a point just to the right of the vertical centre-line of the yo-yo, and frictional force acts on the yo-yo where the centre-line touches the horizontal surface, creating a torque that opposes the anti-clockwise rotation so that the yo-yo rolls to the left rather than the right. When *a* is large and the thread is pointing nearer to the horizontal surface on which the yo-yo is resting, the yo-yo still wants to rotate anti-clockwise. However, there is a strong lateral force – the thread – pulling the yo-yo to the right. Again, friction acts to the left of *b* and this time it doesn't just oppose the anti-clockwise motion. Instead, in combination with the lateral force, it turns the yo-yo clockwise and the yo-yo rolls to the right.

Even more intriguingly, between these two extremes, the yo-yo will simply spin without moving one way or the other.

The bumble balloon

Why, when you blow up a balloon and then release it, does it spiral around as the air is expelled rather than following a straight path?

This one is probably best done outside. While experimenting at home the author managed to damage a glass photograph frame and knock over a vase of flowers left over from the 'Flower power' experiment on page 21. Randomly released balloons and their wildly fluctuating flight paths carry more force than you might expect.

What do I need?

- party balloons
- strong lungs

A cup and drinking straws will allow you to investigate the balloon motion more thoroughly, and if you wish to study the more counterintuitive properties of balloons, you'll also need a piece of plastic or rubber tube and a party squeaker.

What do I do? Blow up the balloons and pinch the openings to make sure no air escapes. Hold them above your head and release them.

What will I see? The balloons will fly around in random, circular paths rather than in a straight line.

What's going on? For a balloon to fly in a straight line, the direction of the jet of expelled air would have to be in line with the balloon's centre of mass and its centre of drag – the point where the forces resisting the balloon's forward motion are symmetrical.

If these two centres don't coincide, the centre of drag should be behind the centre of mass, otherwise stability is compromised. The reason darts and arrows have flights is to keep the two centres in line and ensure drag is at the rear of the moving projectile.

If the balloon's line of thrust does not pass through the centre of mass (which is almost certain) but is in the same

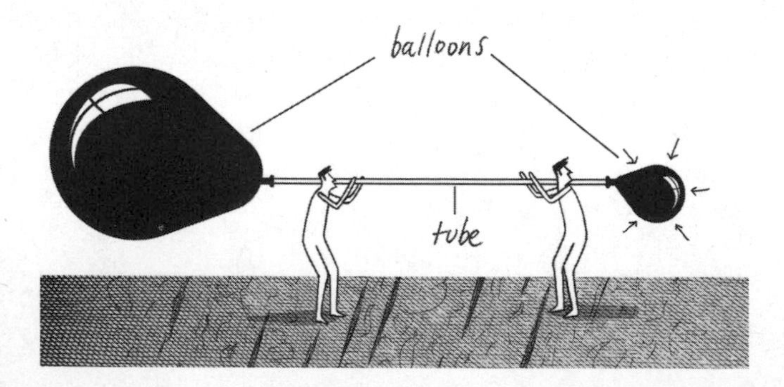

plane as the line joining the centres of mass and drag (which is unlikely), the balloon would travel in a circle in that plane, although the pull of gravity will ultimately force it down to the ground, especially as the air driving it forward expires. However, because these lines generally do not intersect, thrust from the balloon's opening comes at an angle to the plane of the circle, pushing the balloon into the helical, screw-like motion you saw when carrying out the experiment. The thrust of the balloon and the air resistance to the balloon will not cancel each other out in such a situation and so a turning moment is exerted.

To see a similar effect try pushing your two index fingers on the opposite sides of a cup. If the two fingers line up directly at opposite sides of the cup, it will remain stationary. However, if your fingers are not lined up, the cup will turn, in the same way the two forces – thrust and air resistance – acting on the balloon cause it to turn.

Balloons are not finely produced pieces of high-precision engineering, so it is likely that the line of thrust will never be near the centre of drag and the resulting torque will always make the balloon spiral wildly. However, you can aim the balloon to a certain extent by taping a nozzle to it at different angles (try cutting one from lengths of drinking straw). With trial and error, you may be able to get the balloon to fly in a reasonably straight line.

PS: Balloons can teach us quite a bit about physics. Try this: blow up two identical balloons to different sizes and pinch the ends so that the air does not escape. Connect their openings with a piece of plastic or rubber tube using sticky tape and the help of a friend if it proves too difficult on your own. Release the pinched ends so air can flow freely through the tube.

Intuition tells you that the balloons will equalise in size. But they don't. Instead, the small balloon shrinks as it forces its air into the bigger one. A number of theories have been propounded to explain why this happens, and it is likely that all contain some truth. The common theme is that the air moves from the smaller balloon to the larger balloon because the pressure inside a balloon increases as its radius decreases. For a compelling demonstration, stretch the opening of an inflated balloon over a party squeaker. As the balloon deflates, the pitch emanating from the squeaker rises rather than falls, because the air is forced out ever faster. This suggests the pressure in the balloon does indeed rise as it shrinks.

Assuming we accept this, the question is: Why? Some believe it's because balloons act like bubbles, which also display this relationship between pressure and radius. However, balloon rubber does not behave in the same way as a soap film. The tension in the rubber changes in a non-linear fashion – that is, it is not a simple function of how much the rubber has been extended, and common experience supports this. The most obvious example is when you start blowing up a balloon. The initial extension of the rubber requires a lot of puff, and some people actually struggle to get past this stage. Once you do, however, the balloon inflates quite easily. This indicates that the tension in the rubber exerts a higher pressure at the initial low-extension stage than it does when it is much larger, which would account for the smaller balloon expelling its air into the larger balloon.

Up in the air

How do paper planes fly?

We've all made paper planes and some fly quite well, while others crash. Yet even the ones that show good aeronautical ability seem to have no obvious aerodynamic features such as a lift-inducing aircraft wing. Indeed, the ones that fly best have hardly any wing area at all. So how do they fly? Are they like aeroplanes, or is something else at work?

What do I need?

- sheets of A4 paper
- a windless, open space (a sports hall or similar is best, but your back garden on a still day is nearly as good)
- a decent throwing arm

What do I do? Make the paper plane to your usual specification – most people prefer a Concorde-like dart or a delta wing with a long tail, but you probably have your own design – and launch it to see how well it flies. You can tweak the length and size of the wings or the shape of the delta wing. If you've never made a paper plane before, or even if yours is a great flier, try the design below, which combines all the great characteristics of a good paper plane and, we hope, will demonstrate just how a sheet of A4 can become airborne.

What will I see? Your plane will fly – if it doesn't, try our design. You'll notice some loop wildly, others stall, while some fly straight but heavily towards the ground. Patience should eventually produce a version that remains pleasingly aloft for some time.

What's going on? Even a stone has some very limited aerodynamic features if thrown well – consider how far a shot-putter can heave a metal ball. The mechanism by which a paper aeroplane flies can be simply accounted for by Newton's second law: force (in this case, lift) equals the rate of change of momentum. In simpler terms, the plane flies faster and for longer if you throw it harder unless there is something (such as a flapping piece of paper on the nose or dangling, floppy wings) that causes drag and slows down the plane, stymieing its forward motion.

A sheet of paper cannot be thrown as far as a small stone of the same weight because of drag, which is a combination of air resistance and turbulence. Air resistance, caused by the viscosity of air, is proportional to an object's frontal cross-sectional area – or how big the plane's nose is when you look at it front-on. Turbulence is a result of twisting air currents and vortices that form around the plane. It is proportional to surface area and is reduced by a streamlined shape.

In a paper plane, long wings are useless because they are flimsy and cannot resist the bending effect involved in the lifting force. That's why short wings work better – long ones add weight and drag. Weight stops the plane flying far and much of the weight of the plane comes from the wings, but the greatest lift comes from the wing area nearest the centre line of the plane, so the solution is to have very small wings.

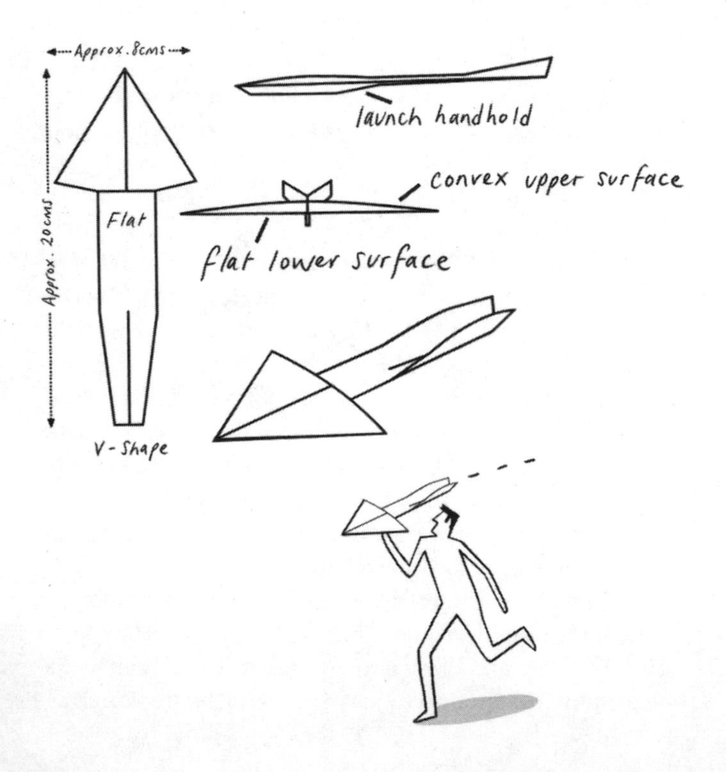

The other key factor to consider is the angle at which the plane sits as it moves forwards. Air needs to strike the underside of the wings and be deflected downwards, ensuring a corresponding upwards force on the plane. If it does not sit at an angle to the flight direction with its nose pointing slightly upwards, it will not generate lift. In a standard paper dart, most of the weight is at the back, which means the back drops, the nose rises and we have lift. This lift balances the weight and the paper dart flies. That's why – unless this lift is too effective, forcing the plane to shoot rapidly upwards and stall – paper darts are better fliers than paper aircraft that actually look like real planes.

PS: When researching what makes paper planes different from real aircraft, *New Scientist* was delighted to hear from reader Mark Wareing, whose design for a paper plane, along with his interesting explanation, follows. In tests in the *New Scientist* office using Mark's design alongside a variety of others, we found that his design stayed aloft for around 25 per cent longer than more familiar paper plane configurations. We also annoyed a lot of people who were trying to work.

The ideal shape is long and thin, tapered towards the front and rear like an arrow, with a small surface area for its given weight. Folding a sheet of paper into this general shape will increase the distance it can be thrown.

In addition, if the paper is folded to produce a flat, horizontal lower surface, lift will support the plane as it descends and increase the distance it can travel. As we noted above, lift is produced as air strikes the underneath of the plane as it glides with a slight nose-up attitude.

In a conventional aeroplane, long, narrow wings near the middle of the body are balanced by small horizontal wings near the tail. Paper is too flimsy for building this kind of structure and flexes, particularly during launch. This means paper wings lose their aerodynamic characteristics and cause drag. And while the traditional paper dart is a streamlined, fairly rigid, delta wing which often produces a stable, gliding flight, it has relatively large wings that produce excess drag and can lack rigidity.

The diagram shows how these problems can be overcome. Smaller, more robust lifting surfaces in the form of a delta wing produced from two or more layers of folded paper retain their shape better during a high-speed launch. They can be folded with a slight convex curve on the upper surface to increase lift in the way a real aeroplane's wings are designed. Additionally, a tightly folded structure adds mass to increase the launch momentum without significantly increasing drag.

If this delta wing is moved to the front and the lift it creates balanced by a long, flat body folded into a V-shape towards the tail to prevent sideways movement (or yawing) during flight, the best characteristics of the paper plane can be combined into one design.

Finally, a flap of folded paper projecting downwards from under the nose of the plane can prove useful as a handhold for launching.

Castles of sand

Why does damp sand make better sand castles than dry sand?

The answer seems obvious: surely water just makes the sand stickier. But there's more to it than that. Try building a few sand castles and see what happens. Obviously, you can try this out at home, but the experiment is much better if carried out on a beach, under a blue sky, with the waves lapping at your toes and a plentiful supply of ice cream.

What do I need?

- sand
- water
- buckets (preferably ones with moulds shaped like castles)
- spades
- ice creams (optional)

What do I do? Make sure that some of the sand is damp (you can try out different proportions of sand/water mix) and some is dry. Pile dry sand into one bucket using your spade and damp sand into another, then pat the sand down into the buckets to level it off. Upturn the buckets, pat them gently on their undersides and remove them from the sand castles.

What will I see? The damp sand castle will retain its shape while the dry sand castle will crumble.

What's going on? To understand this problem you need to consider two ideas: friction and surface tension.

If you place a small block of wood on a table you can slide it around easily, but if you put a heavy weight on top of the block you will need to apply a much greater force to move the block – not just to get the extra mass moving, but also to overcome the increased friction. The bigger the force holding the block against the table, the bigger the force needed to make it slide.

Sand consists of lots of small, hard grains that can slide over each other. If the forces pushing the sand grains together are small, then the grains slide easily over each other. This is the case with the dry sand in the bucket. The sand grains are only pushed together by the weight of any grains above them, so when you tip the bucket upside down they slide easily until they form a conical pile.

When the sand is damp, each grain is coated with a thin film of water, which tends to collect at the points where the grains touch. Surface tension acts at the surface of the water, producing the same result as if the water were covered by a stretched skin that is always in tension. You can see another example of surface tension if you fill a glass of water just a bit too full – a little water can be held in place a millimetre or two above the rim of the glass by surface tension (you can find out more about this kind of surface tension in the 'Floaters' experiment on page 58).

Where the water droplets adhere to the sand grains, the tension is applied to each grain and this effectively pulls each

States of saturation of liquid-bound powders

pendular

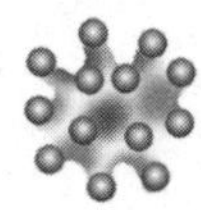

funicular

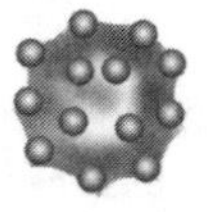

capillary

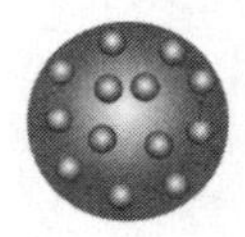

droplet

one against its neighbour, providing quite a strong force between them, even if there is no weight of sand above. The effect of this force, related to capillary action, is enough to provide plenty of friction so that your sand castle stays together.

As you start adding water to the sand, it forms 'pendular' bridges between the sand grains which hold them together, thanks to surface tension and friction. These forces acting together can resist gravity and prevent the walls of the sand castle from collapsing.

The surfaces of these liquid bridges are concave. This generates further capillary action, or suction, which also helps to hold the grains together. If you add a little more water to the sand the pendular bridges start to merge and the sand/water mix passes through the funicular state to reach the capillary state named after this force. In this state the concave liquid surfaces continue to generate a capillary action, which holds the sand grains together.

However, if you keep adding water you reach a point where the surface curvature of the liquid becomes convex rather than concave and the capillary action disappears. This is known as the droplet state. The water no longer creates any attractive force between the particles and the walls of the castle begin to slump and flow as a liquid slurry.

Interestingly, if you try to make your sand castle entirely underwater with completely wet sand, there are no surfaces between water and air in the sand mix, so there are no surface tension forces holding the grains together and the castle will collapse into a conical pile. This proves that it is not simply the water sticking the grains together as you might assume, but is actually a combination of friction, surface tension and capillary action.

PS: These liquid forces are used commercially to agglomerate or granulate many fine powders, such as dishwashing detergents, fertilisers and the powders you find inside

medicine capsules. In these cases, when the water that has been added to these powders evaporates, salts crystallise out of solution to form permanent solid bridges between the grains that continue to hold the agglomerate together, even when dry. You could use a similar approach when building your sand castles, using a concentrated salt solution instead of water to see if the grains making up the castle stick together even after it has dried out. Or, if you've been following our advice and carrying out this experiment on the beach, use seawater, which has a much higher salt content than tap water, and watch what happens when your damp castles dry out.

Nozzle puzzle

Why, when you fill a bucket of water using a hosepipe and hold the nozzle of the pipe underwater and close to the bottom of the bucket, is it seemingly attracted to the bottom of the bucket?

This is a weird feeling: sometimes the nozzle feels as though it is almost being tugged towards the bottom of the bucket. It's a counterintuitive and unexpected response.

What do I need?

- a bucket
- a hosepipe with a nozzle that can control the flow of water (rather than an open-ended piece of hosepipe) attached to the mains water supply

What do I do? Start to fill the bucket via the hosepipe with the tap turned fully on so that the pressure of water coming out of the nozzle is reasonably powerful. Place the nozzle underwater and close to the bottom of the bucket.

What will I see? The nozzle, once you hit the right position, will appear to be drawn to the bottom of the bucket, rather than away from it as the recoil from the jet of water might suggest would happen. You may have to adjust both the nozzle to find a suitable water-stream speed and its position in relation to the bottom of the bucket in order to feel this effect.

What's going on? When the hose filling the bucket is placed so that the end of the nozzle is under the water and near the bottom of the bucket, the force pulling the nozzle towards the bottom can be explained using the Bernoulli principle. This states that the faster a fluid is moving, the lower the pressure it exerts.

When the nozzle is placed near the bottom of the bucket, the restricted area through which the water can flow between the nozzle and the bottom of the bucket causes the speed of the water to increase and produce a region beneath the nozzle which is at a lower pressure than the surrounding, more slowly moving, fluid. This 'sucks' the nozzle downwards.

PS: While this may seem a tricky concept, you can show that moving a fluid lowers pressure by holding a sheet of paper up to your mouth and blowing across its top surface. The moving air causes a drop in pressure above the paper and the paper rises rather than falls, as you might have expected. This helps to explain how aircraft fly. The wings of an aeroplane are shaped to cause a faster flow of air over their upper surfaces than their undersides, and this difference in pressure causes lift.

Overreaction

What causes the extraordinarily explosive reaction when Mentos sweets are mixed with cola or another fizzy drink?

OK, we've saved the biggest and most spectacular until last. Many people are aware of this incredible reaction thanks to viewing the astounding results on the internet. It's a truly awesome sight and one that we simply couldn't leave out … so, stand back and prepare to be amazed.

What do I need?

- a large open space
- a tube of Mentos mints
- a 2-litre bottle of cola or other fizzy drink (preferably of the diet variety)

What do I do? Open the bottle of cola, making sure it is in an open space and well positioned so that it will not fall over. Open the pack of Mentos and make sure they are all dropped into the cola at exactly the same time. This isn't easy – one strategy is to roll a tube of paper so that it holds all the Mentos vertically and fit the paper tube into the bottle neck and release them all at once; another is to place all the Mentos in a test tube or similar slim vessel, cover the neck with paper, tip the test tube upside down over the open neck of the bottle and pull the paper away so the Mentos all fall at once. Whatever method you adopt, you'll have to be able to run quickly…

What will I see? A volcanic eruption of cola squirting vertically out of the bottle. Some reports have recorded a frothy blast of 6 m in height.

What's going on? There is still some debate as to what exactly causes this reaction. Obviously cola is made from – at the most basic level – phosphoric acid, sugar, water and carbon dioxide held in suspension. The initial theory was that the gum arabic and gelatin in Mentos break down the surface tension in the cola which normally constrains any bubbles, allowing the carbon dioxide in suspension in the drink to expand into huge gaseous bubbles and escape very quickly. A similar situation occurs when boiling rice or pasta. You'll notice that these foods will cause a pan to boil over much more readily than if it contained the water alone. This is because similar substances in the rice or pasta disrupt the mesh of water molecules at the surface of the liquid, making it easier for bubbles to form. However, other items dropped into cola – from coins to sugar – also cause it to foam, so while gum arabic may play a role there are almost certainly other factors.

What is not in doubt is that the huge amount of gas created causes a massive increase in pressure inside the bottle, spraying the liquid out in an incredible soda eruption.

However, many scientists now think that the reaction is a physical one, rather than a chemical one. Mentos themselves are covered in tiny pits which act as nucleation sites for bubbles to form in the way that a dusty glass with champagne added will fizz over more readily than a clean glass, because the dust supplies multiple nucleation sites (see 'Over the top' on page 7). Not only that, but Mentos sink, passing through a lot of cola very quickly and allowing pressure to build dramatically.

PS: If you want to avoid getting covered in sticky liquid use a diet version of your drink of choice. Strangely, they seem to work better. While it has been suggested that artificial sweeteners may have something to do with this, nobody yet knows the reason.

More importantly, do not – repeat, DO NOT – attempt to eat Mentos washed down with cola or other fizzy drinks. This would be extremely stupid... and scientists, generally, try not to be stupid.

Want to read more?

www.stevespanglerscience.com/experiment/00000109

For a spectacular video of multiple cola and Mentos eruptions check out www.eepybird.com/exp214.html

Acknowledgements

Most of the entries in this book began life in the popular 'Last Word' column in the weekly *New Scientist* magazine, where they had been described and explained by the magazine's readers – so a big thank-you must go to them for sharing their knowledge with us. Some entries also come from the best-selling forerunners to this volume, *Does Anything Eat Wasps?* and *Why Don't Penguins' Feet Freeze?*, and have been adapted slightly for you to try out at home.

The 'Last Word' column has been running in *New Scientist* for 13 years. Readers pose questions taken from things they've noticed in the world around them and these are answered by other readers – you can check out the full archive at www.newscientist.com/lastword.ns and pose new questions yourself. Send us a new question or answer and you could find yourself in the next book containing the best of the 'Last Word'. We welcome your input at the website address above, via email at lastword@newscientist.com or to our postal address at The Last Word, New Scientist, Lacon House, 84 Theobald's Road, London, WC1X 8NS.

The author would like to offer special thanks to the following, all of whom who contributed to this book in some way, large or small. Apologies to anybody who has been inadvertently omitted.

Anil Ananthaswamy, The Argonne National Laboratory, University of Chicago, USA, J. K. Aronson (Department of Clinical Pharmacology, University of Oxford, UK), John Ashton, Sophy Ashworth, Jamie Au-Yeung, Anna Baddeley,

Steve Ballinger, Byron Barrett, Hugh Bellars, David Bellis, Ralph Berman, Paul Bishop, David Blake, Emma Bland, Peter Bleackley (Astronomy Group, University of Leicester, UK), Ronald Blenkinsop, Jane Blunt, Claire Bowles, Sue Ann Bowling (University of Alaska, Fairbanks, USA), Michael Brooks, Peter Brooks, Mike Brown, Peter Bursztyn, Stephen Burt, Val Byfield, Alan Calverd, J. Neil Cape, Damian Carrington, Brad Carroll (Weber State University, Utah, USA), Julyan Cartwright (Laboratory for Crystallographic Studies in Granada, Spain), Caspar Chater (Centre for Economic Botany, Royal Botanic Gardens, Kew, UK), Stephen Claeys, Robin Clegg (Particle Physics and Astronomy Research Council, Swindon, UK), Chris Collister, Bill Crowther (Aerospace Division, School of Engineering, University of Manchester, UK), Ben Crystall, Glyn Davies, Michael Davies (University of Tasmania, Australia), Allan Deeds, Dave Duncan, Alexandra Dutton, Martin Eastwood, David Edge, Ian Evans, Leopold Faltin, Marina Fernando (Tun Abdul Razak Rubber Research Centre, Hertfordshire, UK), Tony Finn, Richard Fisher, Tony Flury, Tom Foy, Sam Franklin, Yali Friedman, Stephen C. Fry (Institute of Cell and Molecular Biology, University of Edinburgh, UK), Jessica Fryer (vegetarian chef, Canberra, Australia), Mike Gay, Alison George, Stephen Gisselbrecht, Sally Goble, Ken Goldstein, Melanie Green, Ray Hall, Richard Hann, Martin Haswell, David Hauton, Rob Hay, David Henley, Neil Henriksen (The James Young High School, Edinburgh, UK), Jesper Henson (Firefighting Department, Safety and Environment Authority, Zurich, Switzerland), Tom Hering, Angeles Hernández Y Hernández (Andalusian Institute of Earth Sciences, Granada, Spain), Peggy Ho, Richard Horton, Guy Houlsby (Department of Engineering Science, University of Oxford, UK), Shirley Hui, Hugh Hunt, Alan Hutchison, Simon Iveson, Valerie Jamieson, Matti Järvilehto (University of Oulu, Finland), Kate Johnston (Faculty of Biological Sciences, University of Leeds, UK), Nigel Jones, Liz Kane, Roger Kearsey, Gareth Kelly (Head of

Physics, Penglais School, Aberystwyth, UK), Jennifer Kelly, Guy King, James Kingsland, Susan Krafer, Michael Le Page, Learn2Yo-Yo, Meredith Lloyd-Evans, Han Ying Loke, Ben Longstaff, Thomas Lumley, Alistair MacDougall (Institute of Food Research, Norwich), Peter Macgregor, William Madil, D. P. Maitland (Department of Pure and Applied Biology, University of Leeds, UK), Sally Manders, David Mann, Paul Marks, Geoffrey Martin, Sarah Marwick, Geraldine Mathieson, David May (Physics Department, Hind Leys Community College, Shepshed, Leicestershire, UK), Howard Medhurst, D. J. Mela (Institute of Food Research, UK), Peter Milroy, Alison Motluk, Oliver and Dick Nickalls, Jon Noad (Shell International Geological Exploration Team, Netherlands), Conor Nugent, Joyce O'Hare, David Oliveira (Department of Renal Medicine, St George's Hospital Medical School, London, UK), Padraic O'Neile, Emily Owen, Mike Perkin, Andy Phelps, Simon Pierce (University of Insubria, Varese, Italy), Oreste Piro (Mediterranean Institute of Advanced Studies in Majorca, Spain), Chris Quinn, Ian Russell (Interactive Science Ltd, Derbyshire, UK), Simon Scoltock, Alan Scott, Ben Selinger (Department of Chemistry, Australian National University, ACT, Australia), Hillary Shaw, Neil Shirtcliffe (Nottingham Trent University, UK), Colin Siddons, Andrew Smith, Frances R. Smith, Gabriel Souza, Hans Starnberg, Jim Thompson, Penny Thompson, Per Thulin, Lisa Trevithick, Tom Trull (University of Tasmania, Australia), Connie Tse, Johan Uys, Alison Venugoban, Ana Villacampa (Lawrence Livermore National Laboratory, California, USA), Philip Ward and Class 7C (High Storrs School, Sheffield, UK), M. V. Wareing, Alan Watson (Cardiff University, UK), Peter Webber, Tony Weighill, Andy Wells, Richard Williams, Hugh Wolfson, Roger Wong, John Worthington, Paul Wright.

If you are looking for further inspiration for experiments to carry out at home, visit the excellent websites at www.thenakedscientists.com and www.howstuffworks.com, or

check out Rob Beattie's book *101 Incredible Experiments for the Shed Scientist*.

A special thank-you is due to Alun Anderson; Lucy Middleton; Ivan Semeniuk; Jeremy Webb; the production, subbing, art, web, press and marketing teams of *New Scientist*; Frazer Hudson; Jon Richfield; Techniquest, Cardiff, UK; and everybody at Profile Books, including Paul Forty, Andrew Franklin and Ben Usher, for their tireless efforts in the creation of this volume.

Index

Figures in *italics* indicate illustrations